Analytical Techniques in Food Processing

Analytical Techniques in Food Processing

Mukesh Sharma

RANDOM PUBLICATIONS
NEW DELHI (INDIA)

Analytical Techniques in Food Processing

ISBN 978-93-5111-607-3

Published in 2015 in India by

RANDOM PUBLICATIONS

4376-A/4B, Gali Murari Lal, Ansari Road
New Delhi-110 002
Phone : +9111-43580356, 011-23289044, 011-43142548
e-mail: sales@randompublications.com,
info@randompublications.com, randomexports@gmail.com
Reprinted 2018

Type Setting by : Friends Media, Delhi-110089
Digitally Printed at : Replika Press Pvt. Ltd.

Preface

As an introduction to food process engineering, this book describes the scientific principles on which food processing is based and gives some examples of the application of these principles in several food industries. After understanding some of the basic theory, students should study more detailed information about the individual industries and apply the basic principles to their processes.

Food processing dates back to the prehistoric ages when crude processing incorporated slaughtering, various types of cooking, such as over fires, smoking, steaming, oven baking, fermenting, sun drying and preserving with salt. Foods preserved this way were a common part of warriors and sailor's diets up until the introduction of canned food. These crude processing techniques remained essentially the same until the advent of the industrial revolution.

One of the most important challenges facing the country is providing remunerative prices to farmers for their produce without incurring the additional burden of subsidies. This challenge could be addressed if the processing level and value addition of the agricultural raw produce can be enhanced to meet the growing demand for processed foods. Food processing has an important role to play in linking Indian agriculture to consumers in the domestic and international markets.

The role of food processing becomes critical since food production is targeted to double in the next 10 years. With low farmer price realizations and significant wastage in the food supply chain even with the current level of production, only processing of food can secure farmer incomes against a slump in prices as well as reduce wastages. Further, a vibrant food processing industry can be a catalyst for crop diversification.

The book covers processing of wide range of products from fruits vegetables spices cereals pulses oil seeds sugarcane milk meat poultry and fish with traditional processing and preservation and latest technology.

This book is useful for students, teachers and researchers engaged in library and information science research.I would like to thank my team for standing beside me throughout my career and writing this book. My special thanks go to "Random Publications" who have published the book.

– Mukesh Sharma

Contents

1

Consumption of Food Processing Industries

INTRODUCTION

One of the most important challenges facing the country is providing remunerative prices to farmers for their produce without incurring the additional burden of subsidies. This challenge could be addressed if the processing level and value addition of the agricultural raw produce can be enhanced to meet the growing demand for processed foods. Food processing has an important role to play in linking Indian agriculture to consumers in the domestic and international markets.

The role of food processing becomes critical since food production is targeted to double in the next 10 years. With low farmer price realizations and significant wastage in the food supply chain even with the current level of production, only processing of food can secure farmer incomes against a slump in prices as well as reduce wastages. Further, a vibrant food processing industry can be a catalyst for crop diversification. The impact of increased economic growth in agribusiness through food processing can play a significant role in reducing rural poverty and increasing rural incomes. Food processing leads to significant employment generation- not only directly but also indirectly across the supply chain in production and procurement of raw materials and distribution of food products to consumers.

CONDITIONS AND TRENDS OF CONSUMPTION

In feeding stuffs very great quantities of fats disappear but, with a few unimportant exceptions, they always do so as a constituent of some element of the feed and not as such. Thus the amounts of fat that disappear with grain and legumes fed to animals are truly enormous. In food, as in feeding stuffs, very large quantities of fats are consumed incidentally in the ingestion of meats, fish, poultry, milk, and other animal products and in the form of fats and oils naturally contained in cereals, legumes, fruits, vegetables, and other plant products. However, in the human dietary fats and oils are also consumed practically as such, for example as butter, salad oils, and cooking fats.

QUANTITATIVE DATA UNSATISFACTORY

Not only because fats and oils disappear both in the arts and in the dietary in a vast variety of ways, but for other important reasons, the quantitative study of the consumption of fats and oils encounters great difficulties. In the first place, the very concept of consumption is elusive. Presumably one should exclude, in spite of its potential significance, the fat content of large numbers of meat animals that perish from natural causes and from which no attempt is made to recover the fats.

Commercial consumption is a significant concept, but it excludes a large amount of fats in products that do not enter into trade, and a large amount that are incidental components of meats and other products that do enter into trade, for example, tankage and garbage. Wastes in the process of recovery are considerable; hence the fat content of the product treated may be considerably greater than the fats recovered as such plus those incidentally retained. The amounts ingested by animals and by human beings would be worth knowing, but they would exclude large quantities that are wasted in various ways on the farms and in households and public eating places. Furthermore, a distinction may be drawn between intermediate and final consumption. Large amounts of fats are contained in cereals, nuts, milk, oil cake, and other products fed to animals; these are in part used up by the animals, and in part are converted or stored up and appear in dairy products or slaughtered animals. Hence a total of the fats consumed by animals and those derived from animals would contain, in effect, considerable duplication. The difficulties are enhanced by the fact that our statistical information is defective especially in respect to the data on animal fats.

For the United States, the number of animals killed in inspected slaughterhouses is known; the total slaughter at wholesale is reported by the census of manufactures; but only a guess can be made at the number of animals in other uninspected, principally rural, slaughter. Moreover, there is no clear-cut idea of the fat content per average animal slaughtered; the long-time trend seems to be downward, because younger and lighter animals are coming to slaughter in response to the increasing demand for lamb instead of for mutton, for bacon-type hogs instead of for the lard type, and for baby beef instead of for three- to four-year-old steers. To take even the best ratios derived from packing-house practice and apply them to the total wholesale slaughter of cattle, calves, sheep, and hogs as reported in the biennial censuses, and also to the farm and retail slaughter for which the data are officially admitted to be crude estimates, would be little better than intelligent guessing. At the best, there are gaps in the information so large as to make statistical conclusions on several important components, and on fats and oils as a whole, exceedingly unreliable. It would be of interest to know, for each of the several sources of fats:

- The amount wasted for lack of any attempt at recovery;
- The amount ingested;
- The amount wasted in households and public eating places;
- The amount fed to animals; and
- The amount used in industrial production other than food.

We should like these data for fats produced, fats imported, and fats exported. We should like to distinguish between gross and net consumption. Unfortunately, for many of the sources of fats the statistical information is exceedingly defective and it is difficult to reach reliable bases upon which to make the estimates necessary for arriving at supplementary data. We have made attempts to reach a rough approximation to the amounts of fats and oils available for human consumption or industrial use in the United States in recent years, but at several points gaps are so wide and the procedure is so open to objection that we do not feel justified in presenting here even a preliminary approximation, lest in spite of reservations and qualifications it should be taken for more than it would be worth.

In the discussion that follows, therefore, we have incorporated quantitative data only to a very limited extent. The two principal items omitted, fat of dressed meats and milk fats (including butter), are so difficult to estimate that we have not included them; yet their sum probably exceeds the sum of the fats and oils here itemized. They go largely into edible uses, except as milk is fed to animals, though in considerable measure meat fats are wasted in the household or subsequently recovered from household wastes for industrial use. The principal other items omitted are fats incidentally consumed in fish, poultry, grains, vegetables, and fruits. Judging from estimates of Raymond Pearl for the period, the sum of these items is probably less than a billion pounds a year, exclusive of the fat in grain fed to cattle.

FATS AND OILS IN DIET

The function of fats and oils in the diet is mainly to furnish energy to operate the animal machine. In the body, fats are burned as truly as though they were burned in a candle or under a steam boiler, and the end-products of the combustion are the same—carbon dioxide and water. Moreover, the amount of energy they furnish is the same, namely from 9 to 9.4 calories to the gram, whether they are burned within or outside an animal body. (A large calorie, which is the one here used, is the quantity of heat necessary to warm 1 kilogram of water from 0 deg Centigrade to 1 deg Centigrade. A small calorie is the quantity of heat necessary to warm 1 gram of water from 0 deg Centigrade to 1 deg Centigrade.) They yield more energy than the other important classes of foodstuffs, such as carbohydrates and protein (albumin) Conditions and Trends of Production—

Vegetable oils major products), which furnish from 3.8 to 4.2 calories to the gram. Neither fats nor carbohydrates ingested are wholly burned at once unless they are needed for the operation of the animal machine.

If not so needed, carbohydrates are for the most part completely changed in character by conversion into the fat characteristic of the species. This is stored in the body as a reserve against the possibility of a future period of food shortage. Indeed, most of the fat of domesticated animals is produced from starch. Thus the lard of the hogs of the corn belt is mainly derived from the starch of corn (maize).

It has the characteristics of fat normal to the animal, is normally stiff, and hence is especially prized. Unlike ingested carbohydrate, the fat of the food, if it is not at once burned, is changed comparatively little. For the most part it is deposited in but slightly modified form with other fat, made by the animal from carbohydrate or protein, in the storehouses for fat—the adipose tissue under the skin, in and about the viscera, and elsewhere. Therefore, if the food fat differs in its properties from the fat natural to the animal and if there is a great deal of it in the feed, the storage fat formed by the deposition of food fat gives an abnormal character to the fat which is obtained from the animal after slaughter. The dietary fats may be divided into four groups:

- Those consumed as such on the table;
- Those employed in the preparation of foods;
- Those consumed incidentally in the ingestion of meats and other animal products; and
- Those consumed incidentally in cereals, legumes, fruits, and vegetables.

The fats consumed as such on the table in the United States are principally butter, butter substitutes (chiefly coconut and oleo oil), and salad oils (principally cottonseed, corn, and olive oils). The fats and oils used in the preparation of food are butter, lard, lard compounds (for the most part cottonseed oil), cottonseed oil as such, corn oil, peanut oil, butter substitutes, and, in certain types of confectionery, coconut oil and cacao butter. Fats and oils consumed incidentally in the ingestion of meats and dairy products are found principally in milk, cheese, poultry, eggs, beef, mutton and lamb, pork, and fish.

The fat of meats is of course to a material extent not ingested but wasted except in so far as it is recovered in garbage and similar greases. The fats of the cereals, legumes, fruits, and vegetables as a class are important in the aggregate nutritionally, but have little direct commercial importance, since few foodstuffs are purchased on the basis of fat content. Practically all foodstuffs except sugar, water, and certain condiments contain some fat. Fresh fruits and vegetables contain merely traces; nuts, on the other hand, are rich in fats, as are also chocolate products. Oat meal and corn meal that is made without removing the germ are relatively rich in fat; wheat flour is poorer.

A study of the trends of consumption encounters the difficulty that statistics of fat in the diet over a series of decades are not available. Certain inferences seem nevertheless warranted. Trends of consumption may be grouped in two classes according as they are due primarily to changes in habits or primarily to substitutions made by manufacturers of which the consumer may or may not be aware. The two classes of course overlap, for in certain cases both factors play a part. Thus a manufacturer may create a new product which leads to a new food habit or, vice versa, consumer demand may lead to substitution or creation of a new product by the manufacturer.

INFLUENCE OF CHANGES IN FOOD HABITS

Change of food habits may be of two sorts. It may be purely quantitative or it may be qualitative. By a quantitative change is meant increase or decrease in consumption of a given fat rather than the substitution in the diet of one fat for another. By qualitative change is meant primarily a substitution of one fat for another. However, since for physiological reasons the intake of food is practically constant for any individual under any given set of circumstances, a decrease in fat ingestion usually involves either a substitution for it of some other kind of food, or of some other kind of fat. Conversely, an increase usually involves a decrease in consumption of some other kind of food or of fat. Since all fats and oils have very nearly the same food value, one can be substituted for another without change in the volume of food ingested. This statement applies to energy values. Certain fats contain small quantities of chemical substances of unknown chemical nature but of great importance to health. These are known as vitamins or food accessories.

Different fats contain different amounts of them depending upon their origin, method of preparation, and other factors. From the point of view of their vitamin content all fats are not of equal nutritive value; but these are considerations. However, if another food be substituted, the volume of the diet is thereby necessarily increased since no other food has, weight for weight, so great an energy value. It follows that if the replacement of fat in the diet by other foods goes too far the diet must become very bulky if it is still to furnish the same energy value, and this fact sets mechanical limits to the substitution of other foodstuffs for fat in the diet.

Since the caloric requirements of an individual, as stated above, are constant under given conditions, it follows that under these circumstances a reduction in consumption of a fat involves the substitution for it of either an equal amount of some other fat or a corresponding amount of some other foodstuffs. But conditions do not remain constant for any person. He grows old and requires less food. He may become stouter or leaner, then consuming either more or less food. He may change his means of livelihood from one requiring hard manual labour involving high food requirements to a sedentary occupation requiring a low food intake. He may migrate from a very cold climate where food requirements are relatively high to a hot one where they are relatively low. In any nation all these changes are taking place in some individuals in one direction, in others in the opposite direction. For short periods, the result is a reasonably constant consumption.

Over a long period of time this is no longer true. The age distribution of populations changes. If the birth-rate falls and the average span of life lengthens, the proportion of old people who consume little food increases, the proportion of young people who consume much decreases. If it becomes unfashionable to be stout, less food is consumed by the nation as a whole. If the proportion of manual labourers becomes less and that of machine tenders and sedentary workers greater, the per capita consumption of food tends to fall. Since fat is the most concentrated form of food energy, its intake tends to be reduced rather more than that of other forms of food.

In the United States exactly these changes have been taking place for decades; and there is evidence of a corresponding decline in the nutritional use of fats and oils, as part of a general reduction in the per capita food requirements. This reduction is the result of substitution of machine labour for man labour, the decline of outdoor employment in severe winter weather, the improved heating of buildings, the trend to lower average body weights of the people generally, and changes in the age distribution of the population. The population is acquiring more sedentary characteristics, needs less food, and the average per capita requirements of foodstuffs in terms of calories are automatically reduced. Most of these factors are the result of the economic evolution of the country and in essence reflect improvement in the standard of living.

The rise of the standard of living is responsible for certain other changes in food habits. It results in a diversification of the diet with increase in the use of dairy products, fruit, vegetables, and sugar, and decline in the ingestion of cereals and fats. This general statement does not hold for the fat of milk, the use of which is on the increase. Fat consumed incidentally to the ingestion of meats and dairy products has always been heavy. The consumption of milk and poultry fat is increasing, that of beef, sheep, and hog fats is declining. Plain cooking is being replaced by more fancy cooking, which includes more discriminating culinary uses of fats and oils. American practice has departed from the British custom of boiling vegetables without fat. Much of the crackers and bakers' bread consumed now contains shortening.

These facts might seem to suggest increased use of fats and oils. A survey of the entire field, however, suggests that the net result is a lower per capita ingestion of fats in general. The heavy fat rations of hard workers are a thing of the past—fat-backs and sow-belly are no longer staples. We use more butter, but less hog fat. The current public taste for younger and lighter-weight animals represents a substantial reduction in fat. It seems fair to infer that, milk fat and poultry fat aside, the decline in the ingestion of fats applies to animal fats. Over a generation, indeed, it is possible that there may have been an absolute increase in the per capita consumption of vegetable oils with an absolute decline in the per capita consumption of animal fats outside of milk fat.

INFLUENCE OF CHANGES IN FOOD MANUFACTURE

The changes in trends of food-fat consumption are not all by any means due solely to change in habits. Some of those that are quantitatively extremely important are due to substitutions on the initiative of manufacturers of food-fat preparations—for example, the substitution of lard compounds for lard—or to the increasing availability of new types of fat—for example, coconut oil. Some of the possibilities of substitution in food uses have already been touched upon incidentally.

However, since dietary consumption is the major and the higher or premium use of fats, the practice and possibility of substitution require elabouration because they are basic to any consideration of the commodity economics of this important group of raw materials. The substitution of one fat for another, indeed the mere possibility of such substitution, naturally has the greatest influence upon prices.

Given sufficient price inducement, one fat may in some cases displace another with the widest repercussions upon the producer, the farmer, the trader, and the manufacturer. It is one of the purposes of the *Fat and Oil Studies of the Food Research Institute* to present studies from time to time upon far-reaching movements of this general character. That considerations of price, costs of conversion and refining, lack of technological skill, and legislation in the interests of public health and sanitation limit the volume of inedible fats turned to edible uses, has already been pointed out.

But there are other factors as well that limit not merely the diversion of inedible fats to edible uses but also the substitution of one edible fat for another. The most prominent of these are greater or lesser adaptability of different fats to use in the preparation of different foodstuffs, and psychological and sentimental factors that in the food industries play a greater part than in the arts.

CONSUMPTION OF FOOD PRODUCTS

The consumption of food in India was estimated at approximately INR 4,000 bn at 1993-94 prices (INR 7,250 bn at 2001-02 prices) in the year 2001 -02, growing at a CAGR of 7.8per cent (1996-2002). Food consumption has been driven primarily by increase in per capita expenditure on food items. In the period 1996-2002, the per capita expenditure grew by 6 per cent per annum and population increased at the rate of 1.6per cent per annum. Urban food consumption has grown at 9.1 per cent annually and is estimated at INR 1,540 bn (1993-94 prices). Rural food consumption, on the other hand, has grown at

7per cent annually and is estimated at INR 2,550 bn (1993-94 prices). Uttar Pradesh is the largest consumer market for food products followed by Maharashtra, West Bengal, Bihar and Andhra Pradesh. Among the fastest growing markets are the North East states, Gujarat and Andhra Pradesh.

The share of grain-based products is the highest in the Indian consumer's food basket, followed by milk and milk products, vegetables, edible oil and meat products. The growth rates for fruit, vegetables, meat and dairy products is higher than grains and pulses, indicating a shift in consumption pattern. This also implies the need for diversification in agricultural production to match the change in consumption patterns.

CONSUMPTION OF PROCESSED FOOD PRODUCTS

Table. Estimated Total Food Market in 2003-04 (INR bn) at current prices

Total food Consumption*	8,600
Processed foods**	4,600
Primary processed food (includes packed	2,800
packed milk, unbranded edible oil,	
sugar, pulses, spices and salt)	
Value-added foods (includes processed	1,800
juices, jams, pickles, squashes,	
products - ghee, paneer, cheese, butter,	
branded edible oil; breads, biscuits,	
foods; processed meat, poultry and	
and chocolates; alcoholic beverages - beer,	
and malted beverages)	

Table. Level of Processing in Perishable Products

Products		Level of	
	Organi	Unorgan	Total
Fruits and	1.4%	0.8%	2.2%
Milk & milk	13%	22%	35%
Meat and Poultry			
Buffalo	21%	-	21%
Poultry	6%	-	6%
Marine	8%	-	8%

The market size for processed foods for year 2003-04, at current prices is estimated at INR 4,600 bn, which is about 53 per cent of total food consumption. This includes processing in unorganized sector in dairy (Halwais) and grains (chakkis).The market size excluding these segments is about INR 3,300 bn.

The market size of processed foods at factory cost for 2003-04 is estimated at INR 2,100 bn[3] implying about 60 per cent mark-up from factory cost to market price. However, primary processed products constitute as much as 62 per cent of processed foods consumed, with value-added products being the balance 38per cent.

'excluding consumption of alcoholic beverages and out-ofhome consumption

"excluding out-of-home consumption but including alcoholic beverages and processing in unorganized sectors in dairy (halwais) and grain milling (chakkis)

The low share of value added products in food consumption is due to low level of processing across agricultural products.

The key impediments to growth of processed foods include demand-side, supply-side and regulatory issues. The demand side factors include:

- Low per capita income
- Socio-cultural factors such as preference for freshly cooked products as compared to packaged products and availability of low-cost domestic help

The supply side factors constraining the demand for processed foods include:

- Long and fragmented supply chain with independent players engaged in various value-addition activities such as input supplies, cultivation, processing, distribution, storage and retailing (unlike in other major agriculture/food producing countries, where a large number of food companies are integrated across the chain. *e.g.* ADM, Cargill, Tyson Foods)
- High cost of raw material (due to low productivity and poor agronomic practices) and lack of varieties amenable for processing
- Lack of scale
- High cost of packaging
- Lack of infrastructure (cost of power, lack of cold chains, storage, handling and transportation)
- High cost and poor quality of distribution.

CHEMICALS IN FOOD PROCESSING

The same over commitment to potentially risky methods is found in other phases of food production. Complex problems of storage, cleansing, handling, refining, cooking, mixing, heating and packaging are often solved by the use of chemical additives. A half century ago the yellowish tint in freshly milled flour was removed slowly by aging; today it is removed rapidly by oxidizing agents. The miller no longer has to store flour and wait for it to mature. Emulsifiers are employed, partly to "improve" the texture of a product, partly to speed up processing. Their "functional advantages" in ice cream, for example, are that the "mix can be whipped faster, the product from the freezer is dryer and the mix has less tendency to 'churn' in the freezer."

Chemical antioxidants are used to help prevent oils and fats from becoming rancid, a problem that becomes especially pronounced when natural antioxidants are removed during large-scale processing. Chemical preservatives are added to food to permit longer storage in wholesale depots and to extend the "shelf life" of a product in retail outlets. In time chemicals are turned from mere adjuncts of food production into "technological necessities." Their use may result in new machines and facilities, the abandonment of old processing methods and a broad reorientation of technology to meet the new chemical requirements.

The fact that a chemical is technologically useful does not mean that it is biologically desirable. The two phrases are occasionally used interchangeably, as though they were synonymous. Some of the chemicals most widely used in the production of food have been found to be harmful to animals in labouratory tests. Nitrogen trichloride, or agene, was employed for decades as a bleaching and maturing agent in flour before it was discovered, in 1946, that the compound produces "running fits," or hysteria, in dogs.

The use of agene in the bread industry has since been discontinued. Although there is no evidence that agenized flour causes nerve disorders in man, no one can say with certainty that cereal products treated with agene are completely harmless to consumers, especially if eaten day after day for twenty years or more. The toxicant in agenized flour "produces nervous disturbances in monkeys, but not epilepsy in man," observed the late Anton Carlson, one of the most distinguished physiologists of his day. "That is not sufficient for me, because there are many other types of nervous disturbances that may be aggravated or introduced by this chemical."

Another group of chemicals, notably certain types of polyoxyethylene derivatives, were widely used in the United States as emulsifiers before they were banned as potential hazards to consumers. Former F.D.A. Commissioner Charles W. Crawford is reported to have described the agents as "good paint removers." According to data furnished to the Delaney Committee, rats that had been fed two commercial preparations of polyoxyethylene-lb/> derived emulsifiers showed blood in their feces and developed a variety of kidney, abdominal, cardiac and liver abnormalities.

These preparations were used primarily as bread softeners but also found their way into other foods. The polyoxyethylene derivatives constitute a highly suspect family of emulsifying agents, yet some of them are still added to cake mixes, cake icing, frozen desserts, ice cream, candies, dill pickles, fruit and vegetable coatings, vitamin preparations and many food-packaging materials. Similar examples can be culled from other groups of chemical aids. In general, the toxic effects of many commonly used compounds are difficult to determine, but individually and in groups they have aroused suspicion and concern. "The fortification of oils and fats against oxidation and rancidity is another field in which there are serious questions of possible toxicity," warns F. J. Schlink, technical director of Consumers' Research and editor of *Consumer Bulletin*. "The natural antioxidants which delay rancidity are lost during the factory processing of refined oils and fats; then the attempt is made to restore the antioxidation qualities by addition of fat-stabilizing substances. Most of the materials that have antioxidation properties are known to be toxic to some degree."

The term "technological aid" has been used loosely to include food additives that could be dispensed with entirely if a manufacturer improved his processing methods and the quality of his product. The question of whether a chemical, instead of being legitimately employed to furnish distant consumers with a perishable product, was being used rather as a device for masking inferior ingredients and unsanitary methods arose in sharp form early in the century, when sodium benzoate was employed extensively as a preservative in catsup. Harvey Wiley, the first administrator of the federal food and drug law, regarded sodium benzoate as a hazard to consumers.

Supported by data from feeding experiments performed with human beings, Wiley claimed that the preservative causes injury to the digestive system. He was convinced that catsup need not deteriorate readily if clean facilities and proper ingredients are used in preparing it. Challenged on this point by Charles F. Loudon, a canner in Terre Haute, Indiana, Wiley appointed a bacteriologist, Arvill W. Bitting, to work at the Loudon factory and prepare a formula for a stable catsup preparation without artificial preservatives. Bitting was successful. Moreover, after inspecting numerous canning factories and examining many brands of catsup, Bitting concluded that manufacturers who needed sodium benzoate often used inferior products and were careless about sanitation. "The whole spirit and tradition of the pure-food law is against the use of preservatives and substitutes," declared Mrs. Harvey Wiley more than forty years after the Loudon episode. "The intent of the law is to encourage the manufacture of high-grade ingredients, not to try to hide inferiority and cheapness by the use of chemicals and preservatives." If this is so, the spirit of the law has been ignored for decades.

The most up-to-date surveys of chemical additives in foods show that preservatives are added to cheese, margarine, cereal products, jams, jelly and many other processed foods. Uncooked poultry is dipped in solutions of antibiotics. Aside from the damage to public health that may arise from the use of questionable chemicals as preservatives, the longer "shelf life" acquired by some of these foods may cause a loss of valuable nutrients. Many chemical additives in foods perform no technological or nutritional function whatever. They cannot be regarded as materially useful by any stretch of the imagination. These chemicals enhance the colour of a food or make certain products feel soft and newly prepared; in some instances they have been used to replace costly but valuable nutrients with inferior ones. In the least objectionable cases, an additive will deceive the consumer without impairing his health.

The application of an innocuous vegetable dye to some foods often leads the consumer to believe that he is acquiring a better, more wholesome, or tastier product than is actually the case. In the most objectionable cases, a chemical is toxic to a greater or lesser degree. This type of additive not only serves to change the appearance or conceal the nutritive deficiency of a food; it exposes the consumer to a certain amount of damage.

Nitrates and nitrites, for example, are used to impart a pink colour to certain brands of processed meat. Ostensibly, this is their sole function. The addition of these chemicals to meat, especially frankfurters and hamburger meat, would be objectionable if only because of the deception which makes it difficult for the housewife to distinguish between high-quality and low-quality products. But this is not the only deception perpetrated on the consumer. Used together with salt, nitrite compounds definitely extend the "shelf life" of processed meat products.

When nitrites are ingested, they react with hemoglobin in the blood to form methemoglobin and, like carbon monoxide, reduce the hemoglobin's capacity to carry oxygen. An individual who consumes three to four ounces of processed meat containing 200 parts per million of sodium nitrite (a permissible residue) ingests enough of the compound to convert from 1.4 to 5.7 per cent of his hemoglobin to methemoglobin. Ordinarily, this percentage is insignificant. But if the same individual is a heavy smoker and lives in an air-polluted community, the cumulative loss in oxygen-carrying capacity produced by the nitrites in the food and by the carbon monoxide in tobacco smoke and motorcar exhaust can no longer be dismissed as trivial.

Sodium nitrite is highly toxic in relatively small amounts. About four grams of the compound constitutes a lethal dose for adults. Although Lehman regards 200 parts per million of sodium nitrite as a safe residue, he notes that "only a small margin of safety exists between the amount of nitrite that is safe and that which may be dangerous. The margin of safety is even more reduced when the smaller blood volume and the corresponding smaller quantity of hemoglobin in children is taken into account. This has been emphasized in the recent cases of nitrite poisoning in children who consumed weiners and bologna containing nitrite greatly in excess of the 200 ppm permitted by Federal regulations. The application of nitrite to other foods is not to be encouraged."

Emulsifiers have been used in bread not only as softening agents, which can give stale bread the texture of newly baked bread, but as substitutes for nourishing ingredients. "The record of the bread-standards hearings contain evidence of distribution among bakers of advertising material advocating the replacement of fats, oils, eggs and milk by emulsifiers," George L. Prichard, of the Department of Agriculture, told the Delaney Committee. "The use of such products as components of food may work to the disadvantage of our farm economy by displacing farm products normally used. The record indicates that natural food constituents, such as fats and oils, probably will be reduced in many commercial bakery products if bakers are allowed to employ these emulsifiers."

This substitution affects more than 117 the 117 farm 117 economy, however; it also works to the disadvantage of the consumer. To illustrate this, the Delaney Committee report compared the ingredients in two cake batters prepared by the same company eleven years apart; during the interval emulsifiers were added to the product. The first batter, made in 1939, did not contain synthetic emulsifiers; the second, prepared in 1949, did. "On a percentage basis, the cake batter in 1939 contained 13 per cent eggs and 8.6 per cent shortening.

In 1949, the cake batter contained 6.3 per cent eggs and 4.8 per cent shortening, with somewhat less than 0.3 per cent of synthetic emulsifier." The report noted that a "synthetic yellow dye could be added to provide the colour formerly obtainable through the use of eggs. The utilization of synthetic yellow dye in commercial cake was practiced before the war. There are indications that the use of artificial colouring matter is increased when quantities of whole eggs or egg yolks are reduced in commercial cake formulas." To heighten the insult, among the most commonly used emulsifiers in 1949 were the polyoxyethylene monostearates. The yellow dye referred to in the Delaney Committee report was probably FDC Yellow No. 3(Yellow AB), which, until fairly recently, was added to many yellow cake mixes. The dye often contained impurities of a potent carcinogen and its use in foods was forbidden in 1959. For a number of years, however, both the emulsifier and the dye undoubtedly appeared in many brands of cake and reached large numbers of unsuspecting consumers.

Problems of this kind are not likely to disappear unless there is a basic change in the viewpoint of the F.D.A. "Inherited from the Wiley era is a too common misconception that all 'chemicals' are harmful and the related idea that any amount of a 'poison' is harmful," the F.D.A. observes in a brochure on food additives. "The fact is, of course, that chemical additives, or food *additives* as they are now being called, have brought about great improvements in the American food supply. Additives like potassium iodide in salt and vitamins in enriched food products are making an important contribution to the health of our people and yet it is a fact that both iodine and some of the vitamins would be harmful if consumed in excessive amounts. Many similar examples could be given to refute these common misconceptions."

This argument is grossly misleading. Iodine and vitamins are indispensable to human life, but coal-tar dyes and benzoic acid, for example, are not. If we adhered to a well-balanced diet of properly prepared natural foods, iodine and vitamins would never enter the body in toxic amounts. Coal-tar dyes, on the other hand, are suspected of being harmful to man in nearly any amount if consumed repeatedly and the kindest statement that can be made for the presence of benzoic acid in food is that the compound is "safe under the conditions of its intended use."

The formula "safe under the conditions of its intended use" exposes the consumer to serious risks when it opens the door to the use of food additives whose biochemical activity is not understood. Many unexpected problems may arise when such additives appear in food. A particular additive may seem to be relatively harmless to the organs of the body, but it may be carcinogenic on the cellular level of life. Another additive may produce insignificant or controllable effects when studied in isolation; brought into combination with various chemicals in food, however, it may give rise to toxic compounds. The body, in turn, may make a toxic additive more poisonous in the course of changing it during metabolism. "At one time it was generally believed that whenever a toxic substance was absorbed the body was capable of calling upon special mechanisms for detoxifying the toxicant," Lehman observes. "It was believed also that the metabolic pathway that the toxicant followed always proceeded in the direction of the formation of a less toxic compound. Later work on the metabolism of drugs and toxic substances showed that special mechanisms did not exist, but that the toxicant was subject to the same metabolic influences as those which normally operate in the body. The assumption that the metabolic product was less poisonous than the parent substance from which it was derived is also unwarranted simply because in many instances the toxicity of either the original substance and its conversion product is unknown. In other instances the metabolic product is even more poisonous than its parent. The conversion product, heptachlorepoxide, which is two or three times more poisonous than the parent substance, heptachlor [a widely used insecticide], may be cited as an example of this."

Finally, food additives may cause allergic reactions that are likely to go unnoticed for many years. The reactions need not be severe to be harmful. Otto Saphir and his colleagues at the Michael Reese Hospital in Chicago have recently suggested that the development of arteriosclerosis may be promoted by the sensitivity of the body to allergenic compounds, notably certain antibiotics. By using sulfa drugs to produce allergies in forty-two rabbits, the researchers were able in eight months to cause degenerative changes in the arteries of thirty-one of the animals.

These changes closely resembled arteriosclerosis in man. According to a press account of Saphir's report, the rabbits' reactions "were not apparent on the surface." It would be very imprudent to assume that such effects are produced only by chemicals that cause severe or noticeable allergies. If the data of Saphir and his colleagues are applicable to man, additives with even minor allergenic properties cannot be dismissed as harmless. Today more than 3,000 chemicals are used in the production and distribution of commercially prepared food. At least 1,288 are purposely added as preservatives, buffers, emulsifiers, neutralizing agents, sequestrants, stabilizers, anticaking agents, flavouring agents and colouring agents, while from 25 to 30 consist of nutritional supplements, such as potassium iodide and vitamins. The remaining chemicals are "indirect additives"—substances, such as detergents and germicides, that get into food by way of packaging materials and processing equipment. Many chemical additives are natural ingredients, but a large number are not. The artificial substances that appear in food range from simple inorganic chemicals to exotic compounds whose biochemical activity is still largely unknown.

DISTRIBUTION OF FOODS IN INDIA

Food products are sold through over 5 million food and grocery stores in India. The organised food retail market is estimated at INR 20 bn (approximately 0.2 per cent of food expenditure in India).This is in contrast to developed countries, where food distribution is highly consolidated. The majority of food and food products are retailed through neighbourhood kirana stores. The kirana stores focus on dry food products in the absence of infrastructure for cold storage. Bulk of fresh produce is sold by vendors with push carts. Meat, poultry and marine products are primarily sold by small retailers in wet markets. Such produce is associated with low product quality, lack of variety and low hygiene levels.

Internationally, food retailers have played an important role in improving supply chain efficiencies such as developing storage and transportation infrastructure, training supply chain members on food hygiene and standards and providing scientific know how to farmers. The key impediments to growth of organised food retailing in India include lack of infrastructure, technology and capital.

Foreign Direct Investment (FDI) is not allowed in retailing with the exception of cash and carry formats. This restriction is based on the premise that entry of large international players will displace existing employment and reduce bargaining power of farmers. However, there is empirical evidence from other countries (USA, China) which highlights that opening of the retail sector to FDI has enabled employment creation, disintermediation, increased farmer access to information and wider choice and cost savings for the consumer.

EXPORTS OF AGRICULTURAL AND FOOD PRODUCTS

India has 1.5 per cent (INR 360 bn in 2003-04) share of global agricultural exports (approx USD 522 bn), despite its production leadership in agriculture. India's exports primarily constitutes commodity and primary processed items, where price realisations are low. In addition, many products are showing single digit or negative growth. The reasons for India's insignificant share in global trade include supply side factors such as lack of consistency in supply and quality, lack of cost competitiveness, and demand side factors such as non-tariff barriers, short product life cycles and perception of Indian food products in overseas markets.

Most exporters from India lack scale- for example the largest fresh produce exporter records annual sales of about INR 500mn.This has resulted in lack of economies in operations and renders them uncompetitive. Hence, exporters are not able to establish themselves as long-term players in the export market, rely on opportunistic businesses and are consequently, unable to develop technical and managerial expertise. These factors cumulatively lead to low investments by exporters in brand building, quality improvement and brand development.

STATUS OF FOOD DISTRIBUTION IN INDIA

Food accounts for the largest share of consumer spending. Food and food products account for about 53per cent of the value of final private consumption estimated at INR 8600 bn (2003-04 at current prices). Food products are sold through over 5 million food and grocery stores in India. The size of the organised food retail segment in India is estimated at INR 20 bn (approximately 0.2 per cent of food expenditure).This is in contrast to most of the developed countries, where food distribution is highly consolidated.

In countries such as Australia, France, Germany, Denmark and The Netherlands, the share of top the 10 retailers is more than 80per cent (USA ~ 35per cent, China ~ 15per cent).

Most organised retailers in India are regional and use single formats such as convenience stores, supermarkets, hypermarkets or cash and carry. In contrast, most international retailers have multiple format models. Besides reflecting the stage of evolution of retailing in India, this also highlights the capital constraints of existing players in India, restricting their ability to make large investments.

In India, food companies are much larger in size than the organised food retailers in India. In contrast, the retailers are larger in scale, than food companies, in developed countries. Given the channel power of retailers in these markets, food companies have started establishing partnerships with them to develop and test new products, share consumer information and undertake consumer promotions.

KEY DRIVERS

Increasing need for convenience: The Indian consumer visits about eight to ten outlets to purchase various food products which constitute the daily consumption basket. These outlets include neighbourhood kirana stores, bakeries, fruit and vegetable outlets, dairy booths and chakkies (small flour mills). With changing lifestyles, there is growing paucity of time, and convenience in food shopping is emerging as an important driver of growth of one-stop retail formats. Availability of quality retail space: Until the late 1990s, the high cost of real estate meant that organised food retail business models were not financially viable in metropolitan areas. In the last few years, various factors have led to increased availability of real estate for organised retail formats. About 300 malls are at various stages of construction, across metros and mini-metros in the country. The average size of a mall is about 100,000 sq.ft. About 25 malls each are planned in Delhi and Mumbai, primarily in suburban and satellite areas. This will translate into additional retail space of 30 to 40 million sq.ft over the next three to five years.

KEY IMPEDIMENTS

The key impediments to organised food retailing include lack of infrastructure, technology and capital. Lack of infrastructure and technology: Food distribution in India is characterised by a high degree of complexity given dispersed production on the one hand, and variance in demand across locations on the other. There is a compelling requirement for appropriate infrastructure for storage and transportation such as temperature controlled warehouses and vans. However, the limited scale of operations of retailers has restricted their investment capacity in these areas.

The recent entrants who have fuelled growth of organised formats are attempting to address this issue, but given the large requirements of capital, the process of transformation is gradual. Further, entry of organised retailers, who have larger investment capacity, will also address the issue of creation of infrastructure for storage and transportation.

Lack of capital: Organised food retailing is a capital intensive business with a long gestation period (typically 5-7 years). Investment is required in buying/ leasing land, furniture and fixtures, IT systems, quality control systems, vendor development and infrastructure for warehousing, storage and transportation.

The development of organised food retailing in India has been constrained due to lack of capital. Foreign Direct Investment (FDI) is not allowed in retailing with the exception of cash and carry formats.

IMPACT OF ORGANISED RETAILING ON EMPLOYMENT

The restriction of FDI in the retail sector is based on the premise that entry of large international players will displace existing kirana stores and impact employment. However, there is empirical evidence to suggest that development of large retail formats will translate into greater employment as witnessed in other countries such as China, Japan, Singapore and Malaysia. The sector is labour intensive and contributes significantly to employment. The current share of organised retailing and share of retailing in total employment is tabulated below. Higher share of organised retailing has led to higher share of retailing in total employment. In India, the share of employment of organized food retailers is insignificant, on account of the limited size of this segment at present. Traditional kirana outlets employ approximately 20 million people. However, there is significant shadow employment at these outlets, reflecting that the real employment potential of these outlets is far lower.

INDIA'S POSITION IN WORLD TRADE.

Global agricultural exports (including food products) increased from USD 326 billion in 1990 to USD 520 billion in 2003. Strong expansion in food and agri exports in the mid 1990s, was followed by a decline from 1997 to 2000.

This was mainly due to a slump in prices for various agricultural commodities, on the back of demand-supply mismatches. Since 2001,food and agri trade has been growing. The main items in food and agri trade include fruits and vegetables (-USD 75 billion),cereals and cereal-based preparations (-USD 58 billion), marine products (-USD 57 billion), meat and meat-based preparations (-USD 46 billion), milk and milk products (-USD 33 bn) and beverages (-USD 40 billion).Latin America has by far the largest agricultural trade surplus, followed by Oceania. Brazil and Argentina contribute significantly to the Latin American trade surplus. Asia has the largest trade deficit of all regions, primarily on account of Japan. Western and Eastern Europe as well as Africa have a trade deficit too. About half of global trade flows are regionally contained, partly as a result of regional agreements and partly due to the perishable nature of agri-commodities.

There has been a sharp increase in regional trade agreements (RTAs) over the past decade. A total of 259 RTAs have been notified to the WTO by the end of 2002, although only 176 RTAs are operational. An additional 70 RTAs are estimated to be operational, but not yet notified by WTO and about 70 are under negotiation. A well known example of regional trade is in the European Union, where about 75per cent of agricultural trade is within the region. The intra-regional trade figures for other regions are 33per cent (NAFTA),63per cent (Asia) and 15per cent (Latin America and Africa). The group of G-20 countries, a grouping of developing countries led by Brazil, China, India and South Africa have sought further policy reforms and market liberalization on the part of the developed world. At the Geneva meeting of the WTO, the members have agreed on a further reduction of overall trade distorting domestic support, the elimination of all forms of export subsidies and a reduction of import tariffs. The EU already anticipated this by drastically reforming the Common Agricultural Policy (CAP). Besides agricultural policies and trade arrangements, demand-supply trends and the macroeconomic environment also significantly influence trade. Another important factor is the change in currency rates.

EXPORTS

India's exports of agricultural and food products are about INR 360 billion (approx. USD 8 billion) which constitutes about 1.6per cent of total global trade

(USD 520 billion). Exports of agriculture and food products have grown at 15per cent annually, vis a vis total exports which have grown at 19per cent, in the last decade. The share of agricultural exports as a percentage of India's total exports has decreased from 19 per cent to 13 per cent in this period. Excluding non-food agricultural products (such as paper, cotton and jute), tobacco and seeds; marine exports constitute the single largest product category in agricultural and food exports from India followed by, rice oilmeals, wheat, cashew, tea and coffee. Most of the agricultural products are primary processed. Many products have shown negative or single digit growth. The top 15 export categories in food and agricultural (except for non-food items and tobacco) from India are listed below. With the exception of buffalo meat, rice and cashew, India's share in global agricultural and food trade is insignificant. Marine products which have the highest share in Indian exports, have a 2per cent share of global trade in this category. Among the key food exports, only non-basmati rice, wheat and buffalo meat has shown double-digit growth. Marine products, basmati rice and processed fruit juices/vegetables have shown single digit growth. Products like cashew, tea, spices, coffee have shown negative growth.

IMPORTS

Against exports of approximately USD 8 bn India's import of agricultural products are about USD 5 bn which translates into trade surplus of approximately USD 3 bn (Year 2003-04). India has abolished restrictions on imports of many agricultural and food products post liberalisation. However, tariff quotas are maintained on some edible oils, maize and milk powder.

There are restrictions on imports of certain fats, oils of animal origin, and beef, based on GATT Article XX that permits countries to restrict imports on religious grounds. Some products, such as wheat, rye, oats, maize, rice, canary seed and other cereals, continue to be traded by the Food Corporation of India. Despite liberalisation of imports, India's imports are growing at about 1 per cent per annum for the last five years. While exports cover a wide range of products, imports are skewed towards pulses and oilseeds. Pulses and oilseeds are the key food items in India's imports (73per cent share) of food items displaying year on year growth. Increasing production of edible oils in India can reduce India's dependence on imports to a large extent. Another category which has witnessed growth in imports is Scotch whisky. The volume growth is about 13per cent and value growth is about 6per cent in the last decade. However, the growth is driven by imports of bulk Scotch, which is bottled in India and/or blended with Indian liquor by domestic manufacturers. Bottled Scotch imports have not witnessed any growth due to the current level of import duties.

ISSUES HINDERING EXPORTS

India continues to be either absent or at best a marginal player in most of the leading markets for its exports. Indian players have not succeeded in establishing direct linkages with buyers/consumers in importing countries, as a result of which a large proportion of exports are being further processed and re-exported by other countries.

PRODUCT

Marine products - Decline in raw material availability:

- Fragmented base of suppliers
- Lack of availability of technology *e.g.* for tuna fishing
- High duties on imports of additives/flavourings for making value added products
- Anti-dumping issues (such as excess use of antibiotics) in key markets

Rice

- Outdated milling technology (using rubber roll sheller) resulting in high share of broken rice
- Poor quality of seeds/non-availability of certified authentic seeds leading to lack of consistency in grain quality
- Low cost competitiveness due to state taxes and MSP regime (for non-basmati]

Cashew

- Lack of raw material availability - high dependence on imports of raw nuts, Interstate barriers
- On movement of raw nuts
- Purchase tax for exports in some states (*e.g.*4 per cent in Tamil Nadu]
- Lack of mechanization of processing leading to higher production costs
- Lack of established quality parameters for raw cashew

Key Issues Hindering Exports

Tea- Export market viewed as a 'residual market' to sell surplus production.

- High cost of production
- Production focused on CTC rather than orthodox, whereas demand for orthodox tea is growing faster than for CTC

Coffee- Low realizations due to exports restricted to green coffee and no exports in roasted/instant/branded coffee. Export cess of INR 500/MT leading to low profitability for exporters

BUFFALO MEAT

Mango

- Absence of policy on permitting rearing for slaughter which translates into lack of traceability a key requirement in several importing countries
- Inappropriate facilities at existing municipal slaughter houses.

- License for setting up private slaughter house is difficult to obtain.
- Prevalence of various diseases particularly Foot and Mouth Disease (FMD)
- Lack of exportable varieties (high fibre content; inappropriate appearance and texture and large size of stone)
- Lack of post-harvest treatment facilities such as for vapor treatment
- Lack of packhouses from farm to port

Grapes

High cost of setting up vineyards:

- High cost for obtaining certification for exports. For *e.g.*, the cost of EurepGap certificate is
- INR. 75000/farmer (including the cost of construction of separate storage space for fertilizers and pesticides etc)

Dairy Products

Lack of quality monitoring mechanism in the supply chain as required by many importing countrie.

- Opportunistic production of SMP with low exportable surplus
- Few local manufacturing facilities of value added products
- Lack of appropriate packaging technology

Guar Gum

Variation in production from year to year (high share of rain-fed areas). Increasing domestic demand from non-food applications such as textiles, paper, pharmaceuticals, oil well drilling.

UNECONOMIC SCALE OF OPERATIONS

- Lack of consistency in supply and quality
- Lack of cost competitiveness due to statutory charges, intermediation and wastages/losses Inadequate and inappropriate storage and distribution infrastructure

NON TARIFF BARRIERS SHORT PRODUCT LIFE CYCLE LACK OF BRAND IMAGE

Most exporters from India lack scale- for example the largest fresh produce exporter records annual sales of about INR500mn.The low volume translates into lack of economies in operations and makes exports uncompetitive. Hence, exporters are not able to establish themselves as long-term players in the export market, and rely heavily on opportunistic businesses. These factors cumulatively translate into low investments in upgrading skill sets, product innovation, quality improvement and brand building.

- Low economies of scale high cost and low volume.
- Low investment capacity in product innovation and Brand/Market development.
- Low competitiveness in global markets unable to establish as a long-term player.
- Low efficiencies technological and managerial.
- Countries across the world have successfully addressed these issues to become globally competitive.

RICE, CASHEW AND COFFEE EXPORTS FROM VIETNAM

Rice Increased competitiveness through higher productivity: High competitiveness in rice has been achieved on the back of consistent increase in area and productivity for paddy. Cropping patterns have been adjusted to increase land area for planting winter-spring paddy and summer-autumn paddy (from 2.1 and 1.2 million ha in 1990 to 2.89 and 2.35 million ha in 2000 respectively) and reducing acreage under low - yielding winter paddy (from 2.74 to 2.4 million ha).

Intensive farming and more advanced technologies have led to a consistent increase in paddy yields from 3.69 tons/ha in 1995 to 4.6 tons/ha in 2003; higher than other Asian countries including India (2 tons/ha),Thailand (2.2 tons/ha), Myanmar (3.2 tons/ha) and the Philippines (2.89 tons/ha).

These measures have facilitated the increase in output from 19.2 million tons in 1990 to 31.3 million tons in 1999, and further to 34.4 million tons in 2003. Cashew - Mechanization of processing technology: Vietnam is the largest cashew producer in South East Asia and the third largest cashew exporter in the world after India and Brazil. Over 90per cent of production is exported. The number of cashew exporters have increased from 16 in 1997, to over 50 in 2003.

The cashew processing industry in Vietnam has made significant contributions to enhance exports. The developments in processing technology have enabled Vietnam to export cashew in processed form. Vietnam has developed "Cover split technology-designed by Vietnamese technicians. This technology is cheap and is able to generate a higher ratio of whole seed. Due to easy availability of efficient technology, the number of processing companies increased from 6 in 1986 to 30 in 1994 (with total capacity of 75000 tons/year) and to 62 in 1999 (with total capacity of 250000 tons/year) to about 120 in 2003. Coffee - Government Support: Vietnam is ranked as the second largest coffee exporter in the world since 2000, next to Brazil. It exports coffee to about 62 countries. The two largest markets include the European Union (47 per cent share) and the USA (15 per cent share).The Government has played an important role in increasing coffee exports. Some of the initiatives include:

Inviting foreign investment in production and trading of coffee - many of the world's largest trading houses in the world such as ED and F Man, Newman Groupe and O Lam are present in Vietnam. Promoted application of international standards to domestic coffee. In order to cope with the market changes, the Government and industry bodies carried out studies to determine the appropriate proportion of Robusta and Arabica to be cultivated in line with global demand. Accordingly, Robusta plantations were replaced by Arabica.The aim is to achieve a proportion of Arabica and Robusta as 75:25,to align it with global demand.

EMPLOYMENT GENERATION POTENTIAL OF FOOD PROCESSING INDUSTRIES

Food Processing has significant potential for employment generation not only directly but across the supply chain in production of raw materials, storage of produce and finished products and distribution of food products. For *e.g.* a grant of INR 66.7 million (total investment of approximately INR 250 to 300 million) to 35 units in UP in 2003-04 has resulted in direct employment of 2,500 and indirect employment of 20,000, with a significant rural component. Employment intensity is significantly higher in the Small Scale Industries (SSI) sector as compared to the organized sector for the same level of investment.

The incremental employment in the organized sector in FPI sector by 2015 on the basis of the stated vision is estimated at 8.2 Million. The sectoral break-down is as follows.

The employment intensity in the organised sector is 1.8 direct and 6.4 indirect per million of investment. The ratio of indirect to direct employment is therefore 3.5. Rabo India has estimated investment required in the organised sector of FPI as INR 997 bn in next ten years. Hence the employment generation potential in the orgnaised sector is 8.2 million including 1.8 million direct and 6.4 million indirect for an investment of INR 997 bn. The unorganized sector in food processing requires an investment of about INR 100 bn in next ten years (estimated on the basis of output ratio as 2:1 and capital intensity ratio as 5:1 of organized and unorganized sector. The employment intensity is estimated to be approximately 10 direct employment per INR million of investment in the unorganised sector (Source: Dr.JS Bedi Analysis).

This will lead to direct employment creation of 1 million in unorganized sector. The indirect employment generation in the unorganised sector will be about 1 million (assuming ratio of direct to indirect employment as 1:1). The above analysis assumes no replacement of existing employment.

SSI IN FOOD PROCESSING

The SSI Sector accounts for 95per cent of industrial units in the country,40per cent of value added in the manufacturing sector, 34per cent of national exports and 7per cent of Gross Domestic Product (GDP).The SSI sector is the largest employment generator next only to Agriculture.

It has been estimated that an investment of INR 1 million in fixed assets in the small-scale sector generates employment for forty persons and produces more than four million rupees worth of goods or services.

The Food sector is a leading employer within SSI, providing employment to 480,000 persons (13per cent of SSI).

The SSI sector is less capital intensive with a high potential to generate employment.

However, the efficiency of SSI units is impacted by the following:

- Lack of capital/credit
- Inadequate Training technical/managerial
- Tools and technology (traditional and less efficient)
- Limited market knowledge (demand, food standards)
- SSI's role in semi-processing: The organized large-scale sector is focused on processed foods, where SSI cannot compete due to lack of marketing and distribution strengths. However, SSIs can play an important role in procuring from farmers and primary processing of produce to increase shelf life and make it available to processor/ marketers who have access to the final consumer.
- Need for product innovation and branding: There is a strong need to provide necessary training and RandD support to SSIs to promote product innovation. Also, SSIs have limitations in terms of investments on brand development. There is a need to promote public private participation in supporting collective investment by SSIs in branding.

FINANCING NEEDS OF THE AGRICULTURAL AND FOOD SUPPLY CHAIN

The Indian agribusiness supply chain is highly fragmented, with independent players engaged in various value-addition activities such as input supplies, cultivation, processing, distribution, storage and retailing.

The scale of operations of each entity is limited not only due to the structure of the supply chain, but also on account of Government policies as well as due to demand-related factors.

In contrast, internationally, a large number of food companies are integrated across the chain. *e.g.* players such as ADM and Cargill have business interests in agricultural inputs, procurement, transportation and storage of produce, processing and value addition.

The structural complexion of the Indian supply chain thus translates into limited scale of financing as well as higher risk, given the lack of control of each of the players, on the supply chain.

ISSUES WITH FARMER FINANCING

It is important to understand the issues faced in farmer level financing, both from the borrower's as well as the lender's perspective, and develop solutions which address these issues comprehensively.

Farmer's Perspective - Limited Access to Credit

Non-availability of timely credit from organized sources, mainly on account of the processes and procedures of banks/financial institutions. This compels farmers to rely on intermediaries.

- Inadequacy in credit availability-The 'scale of finance' method, stipulated by RBI, caps total credit availability per farmer.
- Inability to offer tangible security/collateral cover to access credit from banks Subsistence farmers, who constitute a large proportion of India's farming community, are unable to benefit from the organized credit system since they cannot offer adequate tangible security. Further, the security offered (land documents) are notional and cannot be enforced by banks.
- The above factors, thus translate into continued reliance on intermediaries for financing. While intermediary financing is timelyas they do not require the level of documentation as banks, the cost of funds is exorbitant, nearly 4 times that of interest rates charged by banks.
- A more efficient agri-supply chain requires that farmers' financing needs are addressed by banks and linkages of farmers with middlemen are scrapped.

Banker's Perspective

- Lack of credit discipline among borrowers Lack of credit discipline, to some extent is infused by the Government's decision to waive outstanding loans of farmers. Policy-related uncertainties translate into a huge risk for banks. Instead of loan waivers, the Government should seek to impose a short-term moratorium on repayments, to enable farmers to overcome the adverse impact of a poor monsoon/ crop failure.
- Lack of tangible security from farmers- as mentioned above, farmers are unable to provide tangible security to financiers.
- Lack of linkages of farmers with processors- In absence of forward linkages with markets, banks run a huge performance risk, as there is greater potential for farmer default.
- Inadequate insurance cover the existing crop insurance schemes are inadequate, and this further increases the risk profile of farmers

- Norms for priority sector lending- As per the current RBI policy on priority sector lending, kharif credit is not accounted for in priority sector lending. This results in a lop-sided focus on extending financing to farmers, favouring crops cultivated in the rabi season.

Policy-related Issues

- Cooperatives Act: There are about 95,000 Primary Agriculture Credit Co-operatives (PACS) and about 100,000 marketing co-operatives. As per the Cooperative Act, cooperatives are allowed to borrow from Cooperative Banks, DCCBs and RRBs. This provision needs to be extended to scheduled banks, in order to enhance flow of credit to farmers.
- State Warehousing Corporations Act: Private sector scheduled banks are constrained from lending to State Warehousing Corporations. Given the sizeable investment required to create grain storage infrastructure, it would be in the interest of the CWC/SWCs that a wider participation of lenders is permitted.
- Restrictions on land holdings: The State Land Holding Laws prohibit consolidation of land holdings.

Financing to farmers with larger land holdings, and thus larger scale of operations, could result in higher credit flows from banks. Further, the Government could also consider consolidation of wastelands and leasing of these to corporates. This would not only enable increase in financing to agriculture, but also aid crop diversification.

CHALLENGE FACING THE FOOD INDUSTRY

The challenge facing the food industry, and this is linked to shifts in the food system more generally, is how to make healthy foods more marketable, and marketable foods appear more healthy. Both questions are capable of eliciting a response from food manufacturers, but to understand the background to this response we need to examine the patterns of resistance which have grown up to recent changes in the food industry, patterns which echo some of the forms of resistance to environmental. The beginnings of what Belasco calls a 'countercuisine' are linked to a number of quite important social changes which, as we have already noted, shift the individual's attention from the sphere of production relations to those of consumption. Three aspects require our attention: the way in which we arrive at our impressions of what foods to avoid and what foods to consume (the quality of food as a commodity in the market); the way in which food 'signifies' and expresses other aspects of our lives (food as self-enhancement); and finally, the way in which our consumption of food is linked to broader questions of ownership and organization in the economy as a whole (food as part of a broader political economy). Almost unnoticed, issues surrounding each of these aspects of the consumption of food have assumed enormous importance in recent years, but close examination reveals that what at first appear to be almost random evidence of social resistance to the modern food system have their roots in fundamental social changes.

These social changes need to be outlined, and their implications examined. It is our contention that too much attention has been devoted to the 'signifying' aspects of food and diet, the semiology of food consumption, by sociologists and anthropologists, to the detriment of a wider understanding of the social transformations implied by the economic and technological changes in the food system.

As we have seen, women's roles have been transformed in a number of ways, some of which are clearly irreversible. Domestic work, including housework, has been opened up to important market forces, and the manufacture and purchase of 'white goods' for the home is now an important activity. At the same time the idea that housework is no longer onerous has served the interests of manufacturers and others. It is far from clear that technology in the home is labour-saving; this assumption, needs to be examined critically. Evidence exists that a human price is exacted for the benefits of speed and convenience which 'white goods' bring.

Housework standards change, and women expect (and are expected) to raise the standards of housework in turn. At the same time, it is likely that women's release from the worst forms of labour drudgery is associated with more time being spent by them on other activities in the home, particularly attention to children. This is not to say that many household tasks are not easier; they are. It is merely to point out that the definition of housework, and with it women's responsibilities in the home, is constantly shifting. It is not immutable. It is also clear that since the 1970s important changes have occurred in popular understanding of the relationship between food and health. Preventative medicine may only be in its infancy, and receive little official encouragement, but for some groups of people in the industrialized world, healthy eating is now considered essential.

Healthy eating in the past depended critically on local custom and diet and, most importantly, income. Today this is still true, but local variations have less importance, and additional factors have made their appearance, which have engaged the attention of a battery of professionals, such as health educators, doctors, consumer groups and alternative therapists. The perceived need to diet to reduce obesity has given rise to a huge consumer market in low-calorie foods, including the controversial 'low-cal' liquid diets and innumerable diet 'systems'. The publication of evidence about heart disease and cancer has led to some marked shifts in the public consumption of certain foods, especially in the United States, where the changes date back to the publication of the

McGovern Report in 1977. Food has become one of the obsessions of our time, associated as it is with good health, longevity and the occurrence of stress. Physical fitness has also become an obsession in some quarters, perhaps also a 'fad'. Evidence from the United States suggests that, while in 1960 only a quarter of adults exercised regularly, by 1980 over half did so.

The point is not that everybody is exercising more, but that more people are, and that health is increasingly linked to diet in many people's minds. This is not a trend likely to be discouraged by the food industry, since it also represents a change with commercial possibilities. Indeed, the giant food firms have jumped on this bandwagon, adapting to each new nutritional recommendation—low sodium, low saturated fats, high fibre—to the point of confusing consumers grappling with labels and elusive media-speak, such as 'lite' and 'natural'.

Changes have also occurred in public perceptions of the two processes: appropriation and substitution. Following the scare over DDT between 1969 and 1972 in the United States, and subsequent publicity about the use of other pesticides, attention has been given to the effects of pesticide residues in food, especially by the London Food Commission in the United Kingdom. Few studies have been undertaken, and even fewer given publicity, about the relative nutritional quality of organically produced and chemically produced food. Pesticide use in the United States rose 500 per cent between 1950 and 1986, but in the latter year one-fifth of United States's crops were lost to pests, the *same percentage as in* 1950! The tactic employed by the food industry in meeting the criticism of environmentalists and whole food enthusiasts, has been that the industry can 'compensate' through industrial processes for the shortcomings of nature.

The industry argued that organic foods were more expensive (ignoring the hidden subsidies provided to chemically produced food), that organic foods could not meet market demand, that food additives increased the palatability of food, and that convenience removed the drudgery from food preparation. By the late 1960s in North America, and perhaps a decade later in the United Kingdom, food manufacturers began to 'reposition' themselves, as well as their food products. Now they were on the side of emancipated women and environmentalists; they offered themselves as accomplices in women's drive for more independence and in the vanguard of responsible, sustainable resource management.

Market researchers have credited much of the improvement in the food industry to their own responsiveness to consumer pressure. In the words of an industry spokesperson: For example, ten years ago the only place people could find additive-free *natural* products were in the comparatively cramped and small premises of health food stores. Today such products, not only much improved and—dare we say it?—engineered to satisfy consumer preferences, are mainstream products in prime placing in major supermarket chains the length and breadth of Britain.

This represents one side of the story, but in practice healthier eating has also presented a challenge to the food industry which it has found difficult to grasp. The problem, in a nutshell, is how to leave out food additives and processing, and still add value to the final product.

The post-war period had seen additives 'substitute' for natural ingredients, while much that was natural had been removed from food. Food was given more value added through packaging, processing and ensuring it stayed 'fresh' longer.

The challenge represented by making 'healthy' foods from 'health' foods was how to make profits by appearing to do less to the product. The industry responded to the challenge, as we have noted, by both market and product differentiation—identifying, and exploiting market opportunities for 'niche' products, such as low-calorie foods, ethnic health foods and fitness-related foods.

Finally, the food industry has been in the forefront of the promotion of food supplements; people can now supplement their inadequate diet from a range of commercially promoted food 'accessories'.

In 1985 over US$3 billion was spent in the United States on vitamin and mineral supplements and over $4.5 billion in supplements to fortified breakfast cereals. This diet supplementation was the inevitable outcome of the way eating had changed in the United States. In 1909, 40 per cent of calories had been provided from fruit, vegetables and grains. By 1976 only 20 per cent of calories came from these sources, the rest from fats and refined sugars. As people ate more calories they ate fewer vitamins and minerals—supplements have helped to fill the vacuum left by changes in the 'affluent diet' of North America.

Another important social change that has accompanied the shifts in diet referred to above is that governments in the industrialized countries have been forced to make some accommodation to the pressures put upon them by food consumers, and interest groups representing consumers. In the United Kingdom this has come about largely as a result of successive food 'scares', such as those over salmonella, listeria and bovine spongiform encephalopathy ('mad cow disease') in the last few years.

The extent to which governments will respond to consumer anxieties is still unclear, but the current debate about public nutrition policy and food legislation represents an excellent illustration of the kind of contradiction to emerge from the development of the modern food system.

QUALITY CONTROL IN FOOD PROCESSING

The maintenance of a well functioning Quality Assurance (QA) programme is essential if a consistent product is to result which meets all required standards. Such a programme should be based on Hazard Analysis and Quality

Analysis Critical Control Point (HACCP and QACCP) systems. HACCP and QACCP are more proactive than traditional approaches to QA/QC activities. The establishment of such programmes is the responsibility of QA personnel but the execution of it involves everyone in the company. To avoid ambiguity regarding responsibility for any QA function, it is important to assign specific HACCP/QACCP accountabilities to responsible persons and groups. A QA programme must consider all activities impacting upon product quality, from raw materials and ingredients used to product handling through distribution channels all the way to the final consumer. In respect of this, Wilson has outlined the following required components of a QA system:

- Raw material control – standard specifications must be adopted for all ingredients which must then be inspected to ensure conformity;
- Process control – all chemical, physical and microbiological hazards as well as quality factors must be identified, critical control points (CCP) must be established, monitored and a record made of any action taken;
- Finished product control–this requires that the finished product be unadulterated, properly labelled and that the integrity of the finished be protected from the environment.

HACCP AND QACCP

All food production activities must be monitored and controlled within the framework of an effective QA programme. The addition of nutrients to a food for the purpose of fortification adds to the control points which have to be considered. Poor manufacturing control leading to excessively high levels of nutrients in the finished product could have important health implications for the consumer if intake of the nutrient reaches the toxic dose.

Conversely, low levels of nutrients in the finished product could render it nutritionally ineffective. This could also have serious health implications if the target population in the fortification programme was at high nutritional risk. Poor manufacturing control could also lead to other quality defects related to interactions of added nutrients with other components of the system. The following steps in the implementation of a quality assurance programme in the production of a fortified food have been outlined by Wilson:

- Product specifications–All specifications for fortificants, food vehicle and any other ingredients must be documented as well as acceptable deviations of these. These include specification of particle size, colour, potency, level of fortification as well as any other requirement which might be deemed necessary.
- Product safety assessment–This involves an assessment of microbiological, chemical and physical hazards for all ingredients and the finished product

- Product analysis–Sampling and testing procedures for all ingredients and the finished product must be explicitly stated.
- Determination of critical and quality control points – Based on first hand knowledge of the total process (including the plant facility, equipment and environment) stages at which inadequate control could lead to unacceptable health risk or adversely affect product quality are identified. The system of controls and actions to be taken at each control point are documented.
- Recall system–A mechanism must be put in place whereby product can be recalled if such action becomes necessary.
- QA audit–Periodic checks are necessary to verify that the QA system is effective and product quality is maintained up to the ultimate consumer.
- Feedback mechanism–Response to consumers and other relevant groups to correct any deficiencies discovered.
- Documentation of QA system–Details of the QA programme used in the production of the fortified food must be readily available to relevant individuals and organizations.

Shortcomings of many fortification programmes in the past have been due to failure to establish an adequate quality assurance programme. Evaluation of the fortification of sugar with vitamin A in Guatemala showed that only 30 per cent of samples tested were fortified at levels within the legal limits. A study of iodine content in iodised salt samples obtained from several plants in India also provided an example of the need for greater control in processing. In the determination of critical and other control points for any process accurate flow diagrams outlining the total process have been used. The construction of an accurate flow diagram for any given process requires first hand knowledge of the processing facility and its environs so that all factors which might be expected to impact on product safety could be identified. Recommendations of the FAO/WHO Expert Technical Consultation on "The Use of HACCP in Food Control' included the following:

- Use of HACCP serves to improve food safety control and should be applied on that basis;
- The elabouration of food safety policies by government and international agencies should use risk analysis as the basis for establishing food safety priorities and for focusing inspection resources. These policies should be implemented through national strategic plans;
- In the post Uruguay Round of GATT, the Codex Alimentarius Commission should recognise the importance of its role in harmonising and establishing food standards, guidelines and recommendations particularly as it relates to safety of food in

international trade. Codex should develop a strategic plan which will include a strengthening of the scientific basis for risk analysis, equivalence and the elabouration of its standards, guidelines and other recommendations and should include specific instructions to the Codex Committees on incorporating HACCP.

ANALYSIS OF VITAMINS AND MINERALS

Analysis of potency of fortificants and of vitamin and mineral content constitute an important component of the overall analytical requirements in QA/QC programmes for fortification processes. Development or selection of appropriate analytical methodologies must be based on consideration of accuracy and precision of measurements, available facilities and equipment, simplicity of procedure and rapidity of determination.

There are also many other experimental methods which have been developed in various labouratories. The Codex Committee on Methods of Analysis and Sampling is working in close cooperation with ISO, AOAC International and IUPAC to recommend protocols for determining the reliability of analytical test methods and results of labouratory analyses. The community bureau of reference of the Commission of European Communities (BCR) has set up a working group to compare analytical methods used in different labouratories in Europe.

Walter summarised the methods most commonly applied to the analysis of vitamins in foods. HPLC is the technique of choice for many of the vitamins because of the reliability of the method, the rapidity of the determination and the often reduced requirement for rigorous preliminary clean up steps. The drawbacks associated with this technology, however, are high equipment and materials costs and its lack of mobility. A comprehensive review of current methods used in the analysis of vitamins was provided by Lumely.

For analysis of vitamin A content in the MSG fortification programme Muhilal et al. developed a quantitative spectrophotometric method. A semi-quantitative method was also described suitable for field testing in the MSG-vitamin A programme. A rapid quantitative method for the determination of vitamin A content in fortified sugar has been described by Aguilar et al. This is based on the Carr-Price procedure. Adaptations of the method, yielding semi-quantitative results, suitable for field testing have also been reported.

The determination of iodine content in iodised salt has traditionally been carried out using a titrimetric method. Recently a number of HPLC based methods have been used to a large extent. One such method described by de Kleijn was used by the Food Inspection Service in the Netherlands. It utilised reversed phase HPLC with uv detection, and had a detection limit of 2.9 ng KI. Modified HPLC procedures involving precolumn derivatisation have been developed to increase the sensitivity of the analysis.

In this way Verma et al. reported a detection limit of 0.5 ng iodide. Another recent report involved a differential pulse polarographic method which did not require a separation or preconcentration step. For qualitative testing in the field, simple test kits based on the reaction of starch with iodine are available. There are important restrictions to the use of the kits which are currently being used. These are reported to be applicable over a restricted pH range.

Additives commonly used in salt such free-flowing agents or stabilisers of added iodine cause the pH to be elevated above the valid range. In such cases, false negative results are obtained. Modification of this procedure to extend the functional range of testing conditions must be investigated. Users of such field kits need to be educated as to their correct use and their limitations.

In iron fortification programmes, quite often measurement of iron content is inadequate. Bioavailability of the nutrients has been determined by measurement of labelled isotopes absorbed from the diet of human subjects or less complex methods based on animal studies such as the haemoglobin repletion method. Simple chemical methods have been used to approximate bioavailability.

There may be analyses other than the measurement of nutrient content or availability necessitated by a fortification programme. These might include the measurement of colour, particle size or moisture content as well as all testing required for the unfortified food. Based on the identification of critical and other control points, the analytical requirements of the overall quality assurance programme can be determined.

ROLE OF LEGISLATION AND FOOD CONTROL

The primary purposes of food legislation are to protect the health of the consumer and to protect the consumer from fraud. In the case of fortified foods, there is a need to ensure that the population is not at risk of receiving toxic doses of any micronutrient. Food laws must also ensure that the target population does not receive nutritionally ineffective levels of micronutrients. Procedures for monitoring premises where fortified foods are prepared, packed, stored or held for sale as well as mechanisms for penalising defaulters must be clearly defined within the food regulations.

In Switzerland samples of fortified products, both local and imported, are taken from the market annually for analysis of vitamin content by two specialised institutes. If vitamin content is found to be outside of established acceptable limits, the official government agency can disallow sale of the product.

Standards for fortified foods and labelling requirements must also be contained within the food regulations. Standards play an important role in the facilitation of trade, both nationally and internationally. When large differences exist between national standards, however, they can forn technical barriers to trade. In light of the Agreement on Technical Barriers to Trade, the development of international standards for fortified foods is an important step in the elimination of technical barriers.

It has been found that food law is managed most effectively in two parts: a basic food act and food regulations. The act itself should set out broad principles while the regulations should contain the detailed provisions governing the different categories of products. Within the regulations should be found lists of approved fortificant compounds and food standards stating the allowed levels of nutrients in the fortified foods.

This organization gives some flexibility to food law as it is much more difficult to have laws amended than to revise regulations. Prompt revision of regulations may become necessary because of new scientific knowledge, changes in new processing technology or emergencies requiring quick action to protect the public health.

With respect to regulations dealing with fortified foods, changes might be prompted as a result of safety evaluations on nutrient compounds or new information regarding the roles and optimal levels of specific micronutrients in the maintenance of good health. Changes in food processing and packaging technologies could be shown to result in a significant reduction in processing and storage losses of micronutrients, thus requiring a revision in the allowed levels of addition of nutrients. In the face of demonstrated micronutrient deficiencies, regulations regarding standards for certain foods and levels of fortification may need to be revised.

2

Food Process Engineering

INTRODUCTION

As an introduction to food process engineering, this book describes the scientific principles on which food processing is based and gives some examples of the application of these principles in several food industries. After understanding some of the basic theory, students should study more detailed information about the individual industries and apply the basic principles to their processes.

For example, after studying heat transfer in this book, the student could seek information on heat transfer in the canning and freezing industries. To supplement the relatively few books on food process engineering, other sources of information are used, for example:

- *Specialist descriptions of particular food industries:* These in general are written from a descriptive point of view and deal only briefly with engineering.
- *Textbooks in chemical and biological process engineering:* These are studies of processing operations but they seldom have any direct reference to food processing. However, the basic unit operations apply equally to all process industries, including the food industry.
- *Engineering handbooks*: These contain considerable data including some information on the properties of food materials.
- *Periodicals:* In these can often be found the most up-to-date information on specialized equipment and processes, and increased basic knowledge of the unit operations.

BASIC PRINCIPLES OF FOOD PROCESS ENGINEERING

The study of process engineering is an attempt to combine all forms of physical processing into a small number of basic operations, which are called unit operations. Food processes may seem bewildering in their diversity, but careful analysis will show that these complicated and differing processes can be broken down into a small number of unit operations. For example, consider heating of which innumerable instances occur in every food industry. There are many reasons for heating and cooling–for example, the baking of bread, the freezing of meat, the tempering of oils.

But in process engineering, the prime considerations are firstly, the extent of the heating or cooling that is required and secondly, the conditions under which this must be accomplished. Thus, this physical process qualifies to be called a unit operation. It is called 'heat transfer'.

The essential concept is therefore to divide physical food processes into basic unit operations, each of which stands alone and depends on coherent physical principles. For example, heat transfer is a unit operation and the fundamental physical principle underlying it is that heat energy will be transferred spontaneously from hotter to colder bodies.

Because of the dependence of the unit operation on a physical principle, or a small group of associated principles, quantitative relationships in the form of mathematical equations can be built to describe them. The equations can be used to follow what is happening in the process, and to control and modify the process if required.

Important unit operations in the food industry are fluid flow, heat transfer, drying, evaporation, contact equilibrium processes (which include distillation, extraction, gas absorption, crystallization, and membrane processes), mechanical separations (which include filtration, centrifugation, sedimentation and sieving), size reduction and mixing.

These unit operations, and in particular the basic principles on which they depend, are the subject of this book, rather than the equipment used or the materials being processed. Two very important laws which all unit operations obey are the laws of conservation of mass and energy.

CONSERVATION OF MASS AND ENERGY

The law of conservation of mass states that mass can neither be created nor destroyed. Thus in a processing plant, the total mass of material entering the plant must equal the total mass of material leaving the plant, less any accumulation left in the plant. If there is no accumulation, then the simple rule holds that "what goes in must come out". Similarly all material entering a unit operation must in due course leave.

For example, if milk is being fed into a centrifuge to separate it into skim milk and cream, under the law of conservation of mass the total number of kilograms of material (milk) entering the centrifuge per minute must equal the total number of kilograms of material (skim milk and cream) that leave the centrifuge per minute. Similarly, the law of conservation of mass applies to each component in the entering materials. For example, considering the butter fat in the milk entering the centrifuge, the weight of butter fat entering the centrifuge per minute must be equal to the weight of butter fat leaving the centrifuge per minute. A similar relationship will hold for the other components, proteins, milk sugars and so on.

The law of conservation of energy states that energy can neither be created nor destroyed. The total energy in the materials entering the processing plant, plus the energy added in the plant, must equal the total energy leaving the plant.

This is a more complex concept than the conservation of mass, as energy can take various forms such as kinetic energy, potential energy, heat energy, chemical energy, electrical energy and so on.

During processing, some of these forms of energy can be converted from one to another. Mechanical energy in a fluid can be converted through friction into heat energy. Chemical energy in food is converted by the human body into mechanical energy.

For example, consider the pasteurizing process for milk, in which milk is pumped through a heat exchanger and is first heated and then cooled. The energy can be considered either over the whole plant or only as it affects the milk. For total plant energy, the balance must include: the conversion in the pump of electrical energy to kinetic and heat energy, the kinetic and potential energies of the milk entering and leaving the plant and the various kinds of energy in the heating and cooling sections,as well as the exiting heat, kinetic and potential energies.

To the food technologist, the energies affecting the product are the most important. In the case of the pasteurizer, the energy affecting the product is the heat energy in the milk. Heat energy is added to the milk by the pump and by the hot water passing through the heat exchanger. Cooling water then removes part of the heat energy and some of the heat energy is also lost to the surroundings.

The heat energy leaving in the milk must equal the heat energy in the milk entering the pasteurizer plus or minus any heat added or taken away in the plant. Heat energy leaving in

milk = initial heat energy
+ heat energy added by pump
+ heat energy added in heating section
– heat energy taken out in cooling section
– heat energy lost to surroundings.

The law of conservation of energy can also apply to part of a process. For example, considering the heating section of the heat exchanger in the pasteurizer, the heat lost by the hot water must be equal to the sum of the heat gained by the milk and the heat lost from the heat exchanger to its surroundings. From these laws of conservation of mass and energy, a balance sheet for materials and for energy can be drawn up at all times for a unit operation. These are called material balances and energy balances.

OVERALL VIEW OF AN ENGINEERING PROCESS

Using a material balance and an energy balance, a food engineering process can be viewed overall or as a series of units. Each unit is a unit operation. The unit operation can be represented by a box as shown in Figure.

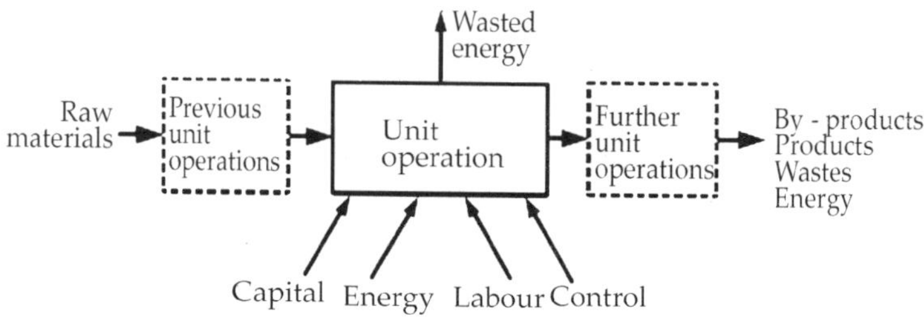

Fig. Unit Operation.

Into the box go the raw materials and energy, out of the box come the desired products, by-products, wastes and energy. The equipment within the box will enable the required changes to be made with as little waste of materials and energy as possible. In other words, the desired products are required to be maximized and the undesired by-products and wastes minimized. Control over the process is exercised by regulating the flow of energy, or of materials, or of both.

TYPES OF PROCESS SITUATIONS

CONTINUOUS PROCESSES

In continuous processes, time also enters into consideration and the balances are related to unit time.

Thus in considering a continuous centrifuge separating whole milk into skim milk and cream, if the material holdup in the centrifuge is constant both in mass and in composition, then the quantities of the components entering and leaving in the different streams in unit time are constant and a mass balance can be written on this basis. Such an analysis assumes that the process is in a steady state, that is flows and quantities held up in vessels do not change with time.

Example: Materials Balance in Continuous Centrifuging of Milk

If 35,000 kg of whole milk containing 4per cent fat is to be separated in a 6 h period into skim milk with 0.45per cent fat and cream with 45per cent fat, what are the flow rates of the two output streams from a continuous centrifuge which accomplishes this separation? Basis 1 hour's flow of whole milk

Mass In

$$\text{Total mass} = \frac{35{,}000}{6} = 5833 \text{ kg.}$$

$$\text{Fat} = 5833 \times 0.04 = 233 \text{ kg.}$$

And so water plus solids-not-fat = 5600 kg.

Mass Out

Let the mass of cream be x kg then its total fat content is $0.45x$. The mass of skim milk is $(5833 - x)$ and its total fat content is $0.0045(5833 - x)$. Material balance on fat:

$$\text{Fat in} = \text{Fat out}$$

$$5833 \times 0.04 = 0.0045(5833 - x) + 0.45x$$

$$\text{and so } x = 465 \text{ kg.}$$

So that the flow of cream is 465 kg h^{-1} and skim milk:

$$(5833 - 465) = 5368 \text{ kg h}^{-1}$$

The time unit has to be considered carefully in continuous processes as normally such processes operate continuously for only part of the total factory time. Usually there are three periods, start up, continuous processing (so-called steady state) and close down, and it is important to decide what material balance is being studied. Also the time interval over which any measurements are taken must be long enough to allow for any slight periodic or chance variation. In some instances a reaction takes place and the material balances have to be adjusted accordingly.

Chemical changes can take place during a process, for example bacteria may be destroyed during heat processing, sugars may combine with amino acids, fats may be hydrolysed and these affect details of the material balance. The total mass of the system will remain the same but the constituent parts may change, for example in browning the sugars may reduce but browning compounds will increase. An example of the growth of microbial cells is given. Details of chemical and biological changes form a whole area for study in themselves, coming under the heading of unit processes or reaction technology.

Example: Materials Balance of Yeast Fermentation

Baker's yeast is to be grown in a continuous fermentation system using a fermenter volume of 20 m^3 in which the flow residence time is 16 h. A 2per cent inoculum containing 1.2 per cent of yeast cells is included in the growth medium. This is then passed to the fermenter, in which the yeast grows with a steady doubling time of 2.9 h. The broth leaving the fermenter then passes to a continuous centrifuge which produces a yeast cream containing 7per cent of yeast, 97per cent of the total yeast in the broth. Calculate the rate of flow of the yeast cream and of the residual broth from the centrifuge. The volume of the fermenter is 20 m^3 and the residence time in this is 16 h so the flow rate through the fermenter must be $20/16 = 1.250\ m^3\ h^{-1}$. Assuming the both to have a density substantially equal to that of water, *i.e.* 1000 kg m^{-3},

$$\text{Mass flow rate of broth} = 1250\ \text{kg h}^{-1}$$

Yeast concentration in the liquid flowing to the fermenter = (concentration in inoculum)/(dilution of inoculum)

$$= (1.2/100)/(100/2)$$

$$= 2.4 \times 10^{-4}\ \text{kg kg}^{-1}$$

Now the yeast mass doubles every 2.9 h, so in 2.9 h, 1 kg becomes 1×2^1 kg (1 generation).

In 16h there are 16/2.9 = 5.6 doubling times
1kg yeast grows to $1 \times 2^{5.6}$ kg = 48.5 kg.
Yeast in broth leaving = $48.5 \times 2.4 \times 10^{-4}$ kg kg^{-1}
Yeast leaving fermenter = initial concentration × growth × flow rate

$$= 2.4 \times 10^{-4} \times 48.5 \times 1250 = 15\ \text{kg h}^{-1}$$

Yeast-free broth flow leaving fermenter = (1250 – 15) = 1235 kg h^{-1}

From the centrifuge flows a (yeast rich) stream with 7per cent yeast, this being 97per cent of the total yeast:

The yeast rich stream is (15 × 0.97) × 100/7 = 208 kg h^{-1}
and the broth (yeast lean) stream is (1250 – 208) = 1042 kg h^{-1}
which contains (15 × 0.03) = 0.45 kg h^{-1} yeast and the yeast concentration in the residual broth = 0.45/1042 = 0.043per cent

Table. Materials balance over the centrifuge per hour

Mass in	(kg)
Yeast-free broth	1235 kg
Yeast	15 kg
Total	1250 kg
Mass out	(kg)
Broth	1042 kg
(Yeast in broth	0.45 kg)
Yeast steam	208 kg
(Yeast in stream	14.55 kg)
	Total 1250 kg

A materials balance, such as in Example for the manufacture of yeast, could be prepared in much greater detail if this were necessary and if the appropriate information were available. Not only broad constituents, such as the yeast, can be balanced as indicated but all the other constituents must also balance.

One constituent is the element carbon: this comes with the yeast inoculum in the medium, which must have a suitable fermentable carbon source, for example it might be sucrose in molasses. The input carbon must then balance the output carbon, which will include the carbon in the outgoing yeast, carbon in the unused medium and also that which was converted to carbon dioxide and which came off as a gas or remained dissolved in the liquid.

Similarly all of the other elements such as nitrogen and phosphorus can be balanced out and calculation of the balance can be used to determine what inputs are necessary knowing the final yeast production that is required and the expected yields. While a formal solution can be set out in terms of a number of simultaneous equations, it can often be easier both to visualize and to calculate if the data are tabulated and calculation proceeds step by step gradually filling out the whole detail.

BLENDING

Another class of situations which arises includes blending problems in which various ingredients are combined in such proportions as to give a product of some desired composition. Complicated examples, in which an optimum or best achievable composition must be sought, need quite elabourate calculation methods, such as linear programming, but simple examples can be solved by straight-forward mass balances.

Example: Blending of Minced Meat

A processing plant is producing minced meat, which must contain 15per cent of fat. If this is to be made up from boneless cow beef with 23per cent of fat and from boneless bull beef with 5per cent of fat, what are the proportions in which these should be mixed?

Let the proportions be A of cow beef to B of bull beef. Then by a mass balance on the fat,

$$\begin{aligned} &\text{Mass in} && \text{Mass out} \\ &A \times 0.23 + B \times 0.05 &=\ & (A + B) \times 0.15. \\ \text{that is}\quad &A(0.23 - 0.15) &=\ & B(0.15 - 0.05). \\ &A(0.08) &=\ & B(0.10). \\ &A/B &=\ & 10/8 \\ \text{or}\quad &A/(A + B) &=\ & 10/18 = 5/9. \end{aligned}$$

i.e. 100 kg of product will have 55.6 kg of cow beef to 44.4 kg of bull beef.

It is possible to solve such a problem formally using algebraic equations and indeed all material balance problems are amenable to algebraic treatment. They reduce to sets of simultaneous equations and if the number of independent equations equals the number of unknowns the equations can be solved. For example, the blending problem above can be solved in this way. If the weights of the constituents are A and B and proportions of fat are a, b blended to give C of composition c:

Then for fat $Aa + Bb = Cc$

and overall $A + B = C$

of which A and B are unknown, and say we require these to make up 100 kg of C then

$A + B = 100$

or $B = 100 - A$

and substituting into the first equation

$Aa + (100 - A)b = 100c$

or $A(a - b) = 100(c - b)$

or $A = 100\,(c - b) / (a - b)$

and taking the numbers from the example

$$A = \frac{100(0.15 - 0.05)}{(0.23 - 0.05)} = \frac{100(0.10)}{(0.18)} = 55.6 \text{ kg}$$

and $B = 44.4$ kg

as before, but the algebraic solution has really added nothing beyond a formula which could be useful if a number of blending operations were under consideration.

HIGH PRESSURE PROCESSING TECHNIQUE

High Pressure Processing (HPP) preserves a food's natural flavour, nutrients, and other sensory properties while extending the shelf life of foods through the inactivation of microorganisms. The uniform application of this non-thermal food processing technology also provides the food industry with new products and new product development opportunities that can fully exploit the functional properties of food ingredients such as hydrocolloids, proteins, etc.

Some of the successful food applications include dressings, sauces, salsas, packaged meats, seafood, cut fruit, purees, juices, chilled ready-to-serve desserts, soups, yogurt, neutraceuticals, pharmaceuticals and other food products. The many advantages of using high pressure processing (HPP) in food production have been known for over a century. However, the technology and equipment required to efficiently and reliability generate the extreme pressures (up to 600 MPa/87,000 psi) used in HPP have only recently become commerciably viable. Food preservation using high pressure is a promising technique in food industry as it offers numerous opportunities for developing new foods with extended shelf life, high nutritional value and excellent organoleptic characteristics. High pressure is an alternative to thermal processing. The resistance of microorganisms to pressure varies considerably depending on the pressure range applied, temperature and treatment duration, and type of microorganism. Generally, Gram-positive bacteria are more resistant to pressure than Gram-negative bacteria, moulds and yeasts; the most resistant are bacterial spores. The nature of the food is also important, as it may contain substances which protect the microorganism from high pressure.

This chapter presents results of studies involving the effect of high pressure on survival of some pathogenic bacteria—*Listeria monocytogenes, Aeromonas hydrophila* and *Enterococcus hirae*—in artificially contaminated cooked ham, ripening hard cheese and fruit juices.

The results indicate that in samples of investigated foods the number of these microorganisms decreased proportionally to the pressure used and the duration of treatment, and the effect of these two factors was statistically significant (level of probability, *P* d" 0.001). *Enterococcus hirae* is much more resistant to high pressure treatment than *L. monocytogenes* and *A. hydrophila*. Mathematical methods were applied, for accurate prediction of the effects of high pressure on microorganisms.

The usefulness of high pressure treatment for inactivation of microorganisms and shelf life extention of meat products was also evaluated. The results obtained show that high pressure treatment extends the shelf life of cooked pork ham and raw smoked pork loin up to 8 weeks, ensuring good micro-biological and sensory quality of the products. Food products are an excellent environment for growth of pathogenic microorganisms, which may cause food-borne diseases.

Quality and shelflife of food products depend greatly on the properties of microorganisms contaminating the food. Despite the introduction of food standards obligatory in EU countries, epidemiologists believe that 75 per cent of food-borne diseases are caused by bacteria.

For this reason, the control of microorganisms is an important aspect of food quality and safety. Many methods of food preservation are used for ensuring microbiological safety, among which high pressure processing (HPP) seems a very promising technique for food industry, as it offers numerous opportunities for developing new shelf life stable foods with extended shelf-life, high nutritional value and excellent organoleptic characteristics—minimally processed but safe for consumers. High pressure is an alternative to thermal processing. The resistance of microorganisms to pressure varies considerably depending on the pressure range applied, temperature and treatment duration, and type of microorganism. As a result of technical progress and government support, the first high pressure processed food products appeared in Japan in the early 1990s.

In Europe, high pressure processing (HPP) of foods was rather at the stage of research or pilot production in the last decade. EU legislation included HPP foods in the "novel food" category. EC Novel Food regulation has introduced a statutory pre-market approval system for novel foods across the whole of the European Union.

Recently, rapid progress of HPP towards commercial exploitation has been achieved, but still the process requires close collabouration between researchers, food and equipment manufacturers, as well as proper financial support.

In Poland, research on the application of high pressure for food preservation was initiated in 1992 by the High Pressure Research Centre (now Institute of High Pressure Physics) of the Polish Academy of Sciences in collabouration with others Institutes.

The research programmes, sponsored by the State Committee for Scientific Research and by EU funds, were devoted to high pressure processing of fruit products, fruit and vegetable juices, milk and meat products. This chapter presents some of the results concerning the effect of high pressure on the survival of *Listeria monocytogenes, Aeromonas hydrophila* and *Enterococcus hirae* in artificially contaminated food products. This microorganisms could be a source of food-borne diseases.

NEED OF HIGH PRESSURE PROCESSING

High Pressure Processing (HPP) is a method of food processing wherein the food is subjected to elevated pressures (pressures up to 87,000 pounds per square inch or approximately 6000 atmospheres) with or without the addition of heat to achieve microbial inactivation or to alter the food attributes in order to achieve consumer-desired qualities.

Pressure is effective in inactivating most of the vegetative bacteria at pressures above 400 MPa. HPP retains food quality, maintains natural freshness, and extends microbiological shelf life. HPP can be used to process both liquid and solid (water-containing) foods.

The process is also called as high hydrostatic pressure processing (HHP) and ultra high-pressure processing (UHP) in the literature. High pressure processing causes minimal changes in the fresh characteristics of foods by eliminating thermal degradation. Compared to thermal processing, HPP results in foods with fresher taste, better appearance, texture and nutrition. High pressure processing can be conducted at ambient or refrigerated temperatures, thereby eliminating thermally induced cooked off-flavours. The technology is especially beneficial for heat sensitive products.

FUNCTION OF HPP

Most processed foods today are heat processed to kill bacteria. Heat oftentimes diminishes the quality of a product. High pressure processing provides an alternative means of killing bacteria which can cause spoilage or food-borne disease without a loss of sensory quality or nutrients. In a typical

HPP process, the product is packaged in a flexible container (pouch or plastic bottle) and is loaded into a high pressure chamber filled with a pressure transmitting (hydraulic) fluid.

The hydraulic fluid in the chamber is pressurized with a pump and this pressure is transmitted through the package into the food itself. Pressure is applied for a specific time, usually 3-5 minutes. The processed product is then removed and stored/distributed in the conventional manner. Because of the uniform manner in which the pressure is transmitted (in all directions simultaneously), food retains its shape, even at extreme pressures. And because no heat is needed, the sensory characteristics of the food are retained without compromising microbial safety.

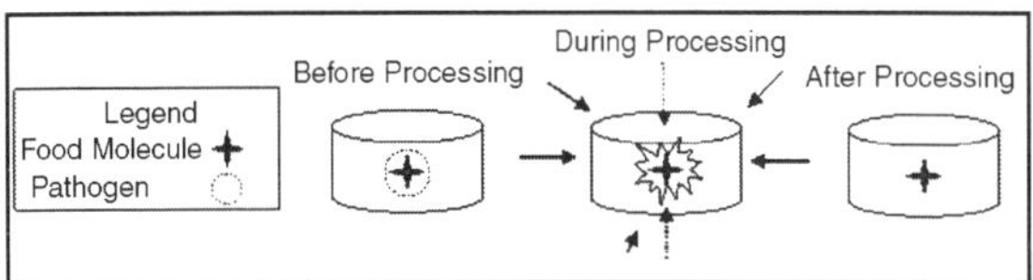

Fig. High Pressure Processing utilizes ultra-high pressures to destroy pathogens without the application of heat that can damage the taste, texture, and nutritional value of the food

HPP AT A GLANCE

- High Pressure Processing (HPP) is based on the Le Chatelier principle which states that actions that have a net volume increase will be retarded and actions that have a net volume decrease will be enhanced.
- HPP utilizes isostatic or hydrostatic pressure which is equal from every direction.
- During HPP, foods are subjected to pressures up to 100,000 psi. which destroy pathagenic microor-ganisms by interrupting their cellular functions.
- Within a living bacteria cell, many pressure sensitive processes such as protein function, enzyme action, and cellular membrane function are impacted by high pressure resulting in the inability of the bacteria to survive. Small macromolecules that are responsible for flavour, odour, and nutrition are typically not changed by pressure.
- HPP is gaining in popularity within the food industry because of its capacity to inactivate pathogenic microorganisms with minimal to no heat treatment, resulting in the almost complete retention of nutritional and sensory characteristics of fresh food without sacrificing shelf life.
- One of the unique advantages of HPP is that pressure transmission is instantaneous and uniform, is not controlled by product size and is effective throughout the entirety of the food item.

- As well, HPP offers several advantages over traditional thermal processing including: reduced process times; minimal heat damage problems; retention of freshness, flavour, texture, and colour; and no vitamin C loss.

Like any other processing method, HPP cannot be universally applied for processing all types of foods. At the moment, HPP is being used in the United States, Europe and Japan on a select variety of high value foods either to extend shelf life or to improve food safety. Some products that are being commercially produced using HPP are cooked ready-to-eat meats, avocado products (guacamole), tomato salsa, applesauce, orange juice and oysters.

HPP cannot yet be used to make shelf-stable versions of low-acid products such as vegetables, milk or soups because of the inability of this process to destroy spores. However, it can be used to extend the refrigerated shelf life of these products and to eliminate the risk of various food-borne pathogens such as *Escherichia coli, Salmonella* and *Listeria*. Acid foods are particularly good candidates for HPP technology.

Another limitation is that the food must contain water and not have internal air pockets. Food materials containing entrapped air such as strawberries or marshmallows would be crushed under high pressure treatment, and dry solids apart from caking do not have sufficient moisture to make HPP effective for microbial destruction.

During HPP processing, pressure is uniformly applied around and throughout the food product. For example, a grape placed between fingers can be easily squeezed and broken; this is because the pressure is not applied evenly from all sides simultaneously. On the other hand, if the same grape is squeezed from all sides simultaneously, it will not be crushed. This can be demonstrated by placing a grape inside soda bottle filled with water. By squeezing the bottle, you pressurize the water inside as well as the grape. Yet the grape is not damaged, no matter how hard you squeeze. In the same way, foods processed by high pressure will not be damaged by the applied pressure.

SHELF LIFE OF HPP PROCESSED PRODUCT

In general, HPP can provide shelf lives similar to thermal pasteurization. Pressure pasteurization kills vegetative bacteria and, unless the product is acidic, it requires refrigerated storage. For foods where thermal pasteurization is not an option (due to flavour, texture or colour changes) HPP can extend the shelf-life by 2-3 fold over a non-pasteurized counterpart, and improve food safety. As commercial products are developed, shelf life can be established based on microbiological and sensory testing.

High pressure processed products are commercially available in the United States, European and Japanese retail markets. Examples of high-pressure processed products commercially available in the US include fruit smoothies, guacamole, ready meals with meat and vegetables, oysters, ham, chicken strips, fruit juices, and salsa. Low acid shelf-stable products such as soups are not commercially available yet because of the limitations in killing spores with HPP. This is a topic of current research.

It has been generally known that high pressure has very little effect on low molecular weight compounds such as flavour compounds, vitamins and pigments compared to thermal processes. Accordingly, the quality of HPP pasteurized food is very similar to that of fresh food products and the quality degradation is influenced more by subsequent storage and distribution rather than the pressure treatment. Pressure also provides unique opportunity to create and control novel food textures in protein based foods.

In some cases, pressure can be used to form protein gels and increase viscosity without using heat. HPP products currently marketed worldwide are primarily distributed refrigerated. In some cases this is necessary for safety (to prevent the growth of spores in low acid foods). For acid foods, refrigeration is not a necessity for microbial stability, but is employed to preserve flavour quality for extended periods of time.

HIGH-PRESSURE FOOD PROCESSING OF RICE AND STARCH FOODS

The use of pressure (P) in addition to temperature (T) has been accepted in the field of food science and technology over the past 15 years, and research and development are now under way in the food industry, universities, and government institutes. Several commercial products are now on the market that are prepared using high-pressure techniques. This section describes the principle of high-pressure treatment and the effects of high pressure on foods, with emphasis on the high-pressure effects on starches. Finally, recent successes of the Echigo Seika Company in the application of high-pressure techniques to rice and rice products are discussed.

Principle and Method

High pressure means high hydrostatic pressure generated by the compression of water. A pressure of 100 MPa or higher (usually lower than 1,000 MPa) is used under temperatures below 100 °C. A unit of pressure is expressed in kg cm^{-2}, bar, or pounds in^{-2}, but now the international Pascal unit is commonly used. Therefore, 1,000 bar or 1,000 kg cm^{-2} almost equals 100 MPa. Food, which is contained in a plastic bag and sealed by carefully removing air, is placed in a pressure vessel filled with water and high pressure is then generated in the vessel by a pressure pump.

Pressurization of Food

An egg is not crushed by compression at 600 MPa, but the egg white and yolk are coagulated. The colour of egg yolk of a pressurized egg is naturally yellow, whereas a boiled egg changes to a faded yellow. The colour difference is attributed to changes associated with pressurization: pressure affects only noncovalent bonding and coagulates proteins without splitting covalent bonds, thus keeping the colour and smell intact. High pressure induces protein denaturation in the same way as high temperature. Meat protein is also denatured by pressure treatment at 400 MPa, preserving the properties of raw meats. An example of prawns or shrimp is interesting: although a boiled prawn turns red and the meat coagulates, the appearance of a pressurized prawn is the same as that of raw shrimp, but the meat coagulates after pressurization at 400 MPa for 10 min.

High pressure also has an effect on starches: a thick suspension of rice starch forms a ball by standing against its own weight after pressurization at 700 MPa, indicating that starches are gelatinized by the pressure treatment.

Versatile Utility of High Pressure

A high-pressure treatment generally coagulates protein, thereby inactivating the enzymes, gelatinizing the starches, and killing microorganisms. Thus, the use of high pressure promises to be a versatile process in food science and technology. Pressure produces a new texture in meats and starch-based foods, while keeping the original nutrients, flavour, and colour.

High-Pressure Effects on Pure Starches

High-pressure-induced Gelatinization of Starch

Effect of Pressure on Amylase Digestibility of Starches: The starches of potato, maize, and wheat are gelatinized by pressure treatment at warm conditions of 45–50°C. The pressurization produces unique properties that are different from those of heat-gelatinization: heat-treatment destroys starch granules, resulting in a transparent solution, but a pressure-treatment swells the granules while maintaining the granular structure.

Nevertheless, amylolytic enzymes such as α-, β-, and glc-amylases digest the pressurized starches well, being similar to the phenomenon in which the pressure-treatment of proteins increases protease digestibility. The pressure-induced gelatinization of starches exhibits a sigmoid curve, suggesting that a two-state transition is involved, as in heat-induced gelatinization.

Effect of Pressurization Time on Amylase Digestibility of Starches: To attain full amylase digestibility of starches, pressurization under warm conditions for 2 to 6 h is necessary. Interestingly, pressurization of starches for a longer time makes amylase digestion difficult: the amylase digestibility of starches decreased by 20–50 per cent after pressurization for 17 h compared with the maximum digestibility obtained after pressurization for 2 to 6 h. These observations suggest that pressure induces gelatinization of starches, similar to heating, but prolonged pressurization produces a new stable structure of starches, which is not susceptible to attack by amylase. *Birefringence of Starches after Pressurization.*

The birefringence of starches is lost as increasing pressure is applied. Wheat starch is sensitive to pressure and birefringence is lost at 200 MPa. In the pressurization of starches, the number of granules exhibiting complete birefringence decreases without increasing the incomplete birefringent granules. This loss of birefringence shows that the crystalline structure is destroyed by high pressure as well as by high temperature and that the high-pressure-induced gelatinization of starches follows a two-state transition without an intermediate state of destruction.

Physical Properties of Pressurized Starches

When a 50 per cent water suspension of starches is pressurized at 100 to 500 MPa and at 45°C for 1 h and air-dried, followed by analyses of their physical properties, the results are as follows.

According to X-ray crystal analysis, the crystalline structure of potato, waxy maize, and maize starches decreases with an increase in pressure, although the crystalline structure of potato starch does not change up to 500 MPa. The decrease in the high-pressure-induced crystalline structure is parallel with the increase in amylase susceptibility. Amylograms show that the transition temperature of pressurized starches is elevated, thus decreasing the viscosity.

Interestingly, differential-scanning calorimetry (DSC) of pressurized starches shows that the peak temperature of the DSC patterns increases while the peak area decreases, indicating that some pressure-induced structural state of the pressurized starches is perturbed by low energy at higher temperatures.

Summary of Pressure-induced Changes in Starches

The structure of pressurized starches changes accompany-ing the increase in amylase digestibility, a loss of birefringence, and a loss of crystallinity. These unique structural properties, which are somewhat different from those of heat-gelatinized starches, should be analysed in detail for effective applications of high pressure to starch and related foods.

Rice-based Foods Produced by High-pressure Processing

Dr. Akira Yamazaki, of Echigo Seika Company Ltd., studied the properties of pressurized rice in detail to introduce the use of high pressure to the food industry. He equipped several large highpressure machines in his own factory and succeeded in sending rice and rice products to the market by introducing the high-pressure technique to the manufacturing process of rice products, cooked rice (*gohan*), rice crackers (*osembei*), and rice cakes (*omochi*).

High-pressure Effects on Rice Grains

Traditionally cooked-rice grains change their shape and develop some cracks, but rice grains after high-pressure pretreatment followed by heat-cooking swell, and their original shape is maintained without cracks.

High-pressure Cooked Rice for Microwave Ovens

Bread is purchased in the store and toasted just before eating. This is a typical life style, especially on a busy morning. However, 45 min are required to cook rice. Cooked rice tastes best and has its best texture just after steaming. Warming of cold cooked rice makes the taste worse. In general, heating a starch-based food twice leads to an unpalatable food. In history, instant-rice is a dream of Japanese consumers.

Dr. Yamazaki succeeded in producing oven-cooked rice with a good taste and texture for consumers, although the production scale is only enough to fulfill the requirement of a small city. He also succeeded in producing instant rice containing miscellaneous cereals.

These instant cooked-rice cereals exhibited good taste and good texture by a 3-min heating in a microwave oven.When high-pressure pretreated and successively heatcooked rice is compared with traditionally cooked rice, its properties are as follows: first, it is more gelatinized; second, it is more slowly retrograded; and third, it is gelatinized to a greater extent by heating just before serving. Japanese consumers who are particularly sensitive to rice accept these properties.

High-pressure Rice Cakes

During New Year days, the Japanese public is freed from the kitchen and enjoys the New Year cerebrations, eating rice cakes (*omochi*) every day, which are a preserved food. The company introduced the high-pressure technique for producing the traditional rice cake and a special kind of *omochi*, which contains herbs with a natural colour, smell, and taste, and this is now available on the market.

Merits of High-pressure Food Processing

High-pressure processing is useful not only for producing high-quality food, but also for improving the manufacturing process. The production time for rice crackers is shortened by introducing a high-pressure technique into the traditional processing system and the total energy cost is decreased by 10 per cent and the labour force by 56 per cent. The use of high pressure for food processing and cooking, in addition to heating and cooling, is now in our hands. Two factors, T and P, are useful for the manufacturing of good foods, including starch-based foods. Although T and P are used independently, their combined use is also important for the optimal use of high pressure. For example, heat-tolerant bacterial spores are inactivated by pressurization under elevated temperature.

INDUSTRIAL APPLICATION OF THE PROCESS FACES ENGINEERING PROBLEMS

So far only applications on a pilot plant scale are reported in the literature. For further developments on a larger scale, theoretical and practical problems should be solved. The industrial application of the process faces engineering problems related to the movement of great volumes of concentrated sugar solutions and to equipment for continuous operations. The use of highly concentrated sugar solutions creates two major problems. The syrup's viscosity is so great that agitation is necessary in order to decrease the resistance to the mass transfer on the solution side. The difference in density between the solution (about 1.3 kg/litre) and fruit and vegetables (about 0.8 kg/litre), makes the product float. Another important aspect, so far not investigated, is the microbiological safety of the process, which should be studied thoroughly before further industrial development.

Osmoappertisation in the Processing of Apricots

In order to obtain an alternative to the canned fruit preserves and to maintain a high quality of the fruits, a research has been carried out on the osmoappertisation of apricots, a "combined" technique that consists in the appertisation of the osmodehydrated apricots. This technique could contributes also to the reduction of energy consumption, limits the cost of production and combines "convenience" (ready-to-eat, medium shelf-life) with many market outlets (retail, catering, bakery, confectionery, semi-finished products).

Apricot Processing

- *Fresh apricot puree:* After washing, cutting and removal of stones, apricot halves are dipped in 2per cent solution of sodium or potassium metabisulphite for 10 minutes. After draining, the resulting material is passed through a 0.045-in. screen pulper - finisher to produce a fresh apricot puree. The fresh apricot puree obtained in this way could be further processed in different semiprocessed (*i.e.* chemically or otherwise preserved products) or finished fruit products (fruit leathers, fruit bars, jams, etc.).
- *Concentrated apricot pulp:* Fresh apricot halves could also be steam blanched for 5 min., passed through a 0.045-in pulper - finisher and transformed in a purée with about 14 Brix depending on the fruit quality. This purée may be concentrated in steam jacketed kettles up to 20 Brix or in other adequate equipment (*e.g.* a stirred vacuum evaporator) up to 28°Brix. As for fresh apricot purée, the concentrate

may be further processed in various semiprocessed or finished fruit products as mentioned above and as will be described below.

- Dried apricot leather
  - From fresh fruit purée by drum drying. The fresh apricot purée at about 14 Brix could be dried to 12per cent moisture apricot sheets, using a double-drum dryer operating at 132 degrees C with a drum clearance of 0.008 in and speed of 45 sec per revolution.
  - From fruit concentrate by drum drying. The concentrate could also be dried to 12per cent moisture fruit sheets by the same process as described above.
  - From fresh fruit purée or from apricot concentrate by sun/solar drying or by dehydration.
- *Trays:* For sun/solar drying or dehydration of fruit pulp, the trays must have a solid base in order to retain the liquid contents. They may be made of metal, timber or plastic. Stainless steel or plastic trays are most suitable because they are unaffected by acid fruit pulp; they are, however, expensive. A metal tray could be 75 x 50 cm in size and with side 5 cm high. The trays must keep level during drying; if the tray is not level the pulp will run to the lowest point, giving a layer of irregular depth which will dry unevenly. Any tray which is not made of stainless steel or plastic must be covered inside with a sheet of heavy gauge plastic film to protect the pulp from chemical or bacteriological contamination. Standard sun/solar trays as described can be used by covering them inside with a sheet of plastic film to create a solid base.
- *Preparation before drying/dehydration*: Fresh apricot purée can be directly used for next processing steps. Fruit concentrate needs to be added to potassium metabisulphite to obtain a 0.3per cent concentration of SO2 in the material.
- *Drying/dehydration*: The apricot/fruit purée or concentrate is poured into the trays to a depth of about 1.5 cm. When stainless steel or plastic trays are used they should be coated with a thin layer of glycerine to prevent sticking.

The pulp is then sun/solar dried or tunnel/cabinet dehydrated; moisture content in the dried product should not exceed 14per cent and the SO_2 content should not be less than 1500 pp. The dried product is wrapped in cellophane to prevent sticking, then put inside polythene bags and stored at best in tight fitting tins and sealed to prevent moisture transfer.From fresh fruit purée or from apricot concentrate, with sugar addition, and then processed by sun/solar drying or by dehydration. In some countries preference is for finished products with added sugar; and this is also interesting from a point of view of energy consumption (concentration is partially achieved by sugar dry matter) and of shelf life. The overall content in SO_2 could also be reduced as sugar is a preservation agent, the product will be close to a fruit "paste".

RECONSTITUTION TEST FOR DRIED/DEHYDRATED PRODUCTS

In reconstitution water is added to the product which is restored to a condition similar to that when it was fresh. This enables the food product to be cooked as if the person was using fresh fruit or vegetable. All vegetables are cooked but many of the dried fruits can be used for eating after they have been soaked in water. The following reconstitution test is used to find out the quality of the dried product. Reconstitution test

- Weigh out a sample of 35 grams from the bulked and packed final product of the previous day's production.
- Put the sample into a small container (beaker) and add 275 ml of cold water (and 3.5 g salt).
- Cover the container (with a watch-glass) and bring the water to the boil.
- Boil GENTLY for 30 minutes.
- Turn out the sample onto a white dish.
- At least two people should then examine the sample for palatability, toughness, flavour and presence or absence of bad flavours. The testers should record their results independently.
- The liquid left in the container should be examined for traces of sand/soil and other foreign matter.

This test can be used also to examine dried products after they have been stored for some time. Evaluation of rehydration ratio may be performed according to the following calculations. Rehydration ratio. If weight of the dried sample is 10 g (Wd) and the weight of the sample after rehydration is 60 g (Wr), rehydration ratio is:

$$\frac{Wr}{Wd} = \frac{60}{10} = \frac{6}{1}, 6 \text{ to } 1$$

Rehydration coefficient. The weight of rehydrated sample is 60 g (Wr); the weight of dried sample is 10 g (Wd) and its moisture is 5per cent (Wu); raw material before drying had 87per cent water (A); rehydration coefficient is:

$$\frac{Wr}{\frac{Wd - Wu \times 100}{100 - A}}$$

$$= \frac{60 \times (100 - 87)}{10 - [10 \times 0.05]} = \frac{780}{9.5} = 52.1.$$

A simpler test for eating quality can be carried out without weighing and measuring. The material is placed in a cooking pot with water (and a little salt).

The pot is then covered and boiled as described above. Except for a few products which are eaten in the dry state, most dried fruit and all dried vegetables are prepared by soaking and cooking. Often this preparation is carried out incorrectly and dried products get a bad reputation. Good quality dried products, after cooking and if properly treated should be similar to cooked fresh produce. In order to get good results, the following methods are recommended:

Quick Method

Cold water, ten times the weight of the dry product, is added to the dried product. The container is covered, brought to the boil and simmered GENTLY until the product is tender. The cooking time may be 15 to 45 minutes after the boiling point has been reached.

Slow Method

This gives better results than the quick method. Cold water is added to the dry food and is left to soak for 1 to 2 hours before cooking. The product is then cooked in the same water as that in which it was soaked. The actual cooking time will probably be shorter than that for the quick method. Other points to remember are:

- If too much water is added the cooked product will have little flavour. However, if too little water is added the product may dry and burn. This can be avoided by adding small quantities of water during cooking;
- Always cook with a lid on the container;
- Salt, if required, should be added when the cooking is almost complete;
- Partly used packages of dry products should be reclosed tightly or kept in containers with good fitting lids.

3

Finance for Agriculture and Food Processing

FINANCING OF THE FOOD PROCESSING SECTOR

The structural complexion of the Indian supply chain has led to limited scale of financing as well as higher risk, given the lack of control of each of the players, on the supply chain.

Financing of farmers: Farmers often rely on unorganized sources of credit due to bottlenecks in access, timeliness in availability and adequacy of credit from organized sources. The key hurdles faced by banks in financing farmers are their inability to provide adequate collateral as security, and the potential for default, in the absence of an assured market for their produce.

Financing of food processing enterprises: The food processing enterprises primarily comprise small and medium sized companies, a large proportion of which have stand alone operations, with no linkages with farmers, and reliant on other organizations to undertake marketing/further processing of their products. Consequently, companies in the food processing sector usually bear a steep cost of interest for the high risk perception associated with the nature of their operations.

Further there are several regulations which limit availability of finance to the sector, including the definition of priority sector lending, the Cooperatives Act and the Warehousing Corporation Act.

Absence of well-developed risk mitigation tools has further impacted availability of finance to the food and agriculture sector. Additionally, crop insurance schemes in India have faced various issues including assessment of farmer yields at the mandal level, lack of past statistical data to calculate premia, high premia for certain crops and administrative hurdles in managing claims, thus resulting in poor economic viability as also limited benefit to farmers.

FDI in food processing: FDI in the food processing sector is low, constituting a mere 4per cent of total FDI in the period 1991 -2004.The actual inflow is only about 28per cent of FDI approved (INR 4.2 bn out of approved INR 11.6 bn) This is despite the fact that with the exception of food retailing, plantations and alcoholic beverages, 100per cent FDI is permitted in the sector.

Government Schemes: Government provides assistance to the industry and entrepreneurs under various schemes. While there are significant overlaps in these on the one hand, there are several unaddressed need gaps of the industry, on the other.

Further, the restriction on quantum of financial assistance per unit disincentivises scale. Additionally, the structure of the scheme needs to be based on specific sectoral requirements. The Government needs to have a robust monitoring mechanism to assess the direct and indirect impact of such assistance.

RESEARCH AND DEVELOPMENT

The Indian food industry needs world class technologies to compete effectively with international products. Indian RandD institutions have not been able to develop innovative products, processes and machinery of global stature as reflected in India's share in global trade.

The industry is dissatisfied with the quality of research output and response of existing RandD institutions. The key reasons for this are segregation of academics from applied research, inadequate industry interface, low commercial orientation and lack of collabourative efforts with global peers.

SOURCES OF FINANCE FOR AGRICULTURE

At present nearly 60per cent of rural credit is used for agriculture and allied activities, 10per cent for non-farm activities and the balance 30per cent for funding household consumption expenditure. Nearly 40 - 45per cent of rural credit needs are catered to by formal credit institutions and the balance by the informal sector including commission agents, input suppliers, traders, Self Help Groups (Joint Liability Groups), processing industries and professional money lenders.

The Indian commercial banks are mandated by the Reserve Bank of India (RBI), to undertake directed lending for agriculture and rural development. Commercial banks are required to achieve priority sector lending of 40per cent of net bank credit of which 18per cent should be for agriculture. Further, sub-targets are also specified *i.e.*, Direct Agriculture (credit for direct farmer benefit, Target: Minimum of 13.5per cent of the net bank credit) and Indirect Agriculture (Target: Maximum of 4.5per cent of the net bank credit).

In FY 03, total agricultural advances outstanding under priority sector lending, by Public Sector Banks stood at INR 735 billion, on a total net bank credit portfolio of INR 4779 billion. As compared to this, agricultural advances by Private Sector Banks stood at INR 119 billion, on a total net bank credit portfolio of INR 718 billion. The following Exhibit provides details of direct and indirect priority sector credit amongst Public and Private sector banks during FY03. In case of a shortfall in lending to agriculture, banks are required to invest in Rural Infrastructure Development Fund (RIDF) deposits, which offers interest rates linked to the bank's performance in lending to agriculture. These rates are inversely proportional to the shortfall in agricultural lending. The interest rate on RIDF is provided in slabs, based on the achievement of the priority sector lending target, and can be as low as 4per cent to 5per cent.Therefore, it is imperative to achieve the agriculture target and more specifically the Direct Agriculture target as it accounts for 75per cent of the total. Established in 1996, RIDF is being utilized for making investments in rural infrastructure projects. Nine tranches of RIDF have been established with an aggregate corpus of INR 340 billion, with the ninth tranche of INR 55 billion. Cumulative sanctions and disbursements under various tranches of RIDF stood at INR 334 billion and INR 193 billion respectively, at the end of February 2004.The RIDF funds are used to finance minor irrigation projects, shallow wells, drinking water facilities, roads, bridges and storage infrastructure.

SOURCE OF FINANCE AND FOOD PROCESSING

Banks: The Food Processing sector has access to credit from Commercial Banks- Indian and Foreign, Cooperative Banks and the Regional Rural Banks. The formal institutional sector assists the food processing companies through long term loans for capital investments and short term loans for working capital.

The total lending to the food processing sector is clubbed with lending to the agriculture sector and separate details for financing to the food processing sector are not available.

- National Bank for Agricultural and Rural Development (NABARD): NABARD offers refinance facilities for food processing, agri infrastructure, developmental assistance for RRBs, DCCBs, SSI and SHG linkages, and assists in infrastructure development and research.
- Small Industries Development Bank of India (SIDBI):SIDBI has been involved in assisting the entire SSI Sector including tiny, village and cottage industries through suitable schemes tailored to address the funding requirements for expansion, diversification, modernisation and rehabilitation.
- Export Import Bank (EXIM Bank): Exim Bank assists in financing and facilitation of foreign trade. The Bank also offers financial support to companies engaged in exports.
- National Cooperative Development Corporation (NCDC): NCDC assists in promotion, planning and financing the agricultural supply chain from production, processing, storage and trade of agricultural produce and food products. NCDC also provides assistance for marketing of certain notified commodities *e.g.* fertilizers, pesticides, agricultural machinery etc.

- Ministries/Government bodies.
- Ministry of Food Processing Industries (MFPI): MFPI is the Nodal agency for development of the processed food sector in the country MFPI's financial schemes include schemes for technology upgradation, Human resource development, Quality testing, RandD,TQM, backward and forward integration, development of infrastructure including food parks.
- Agricultural and Processed Foods Products Export Development Authority (APEDA): APEDA facilitates market linkages between Indian producers, manufacturers and the international market. APEDA provides financial assistance for market development, infrastructure development and development of quality enhancing facilities.
- Ministry of Agriculture (MoA), Government of India: The Ministry of Agriculture under various schemes, provides financial assistance for development of specific crops for investment in seeds, irrigation, farm implements, inputs, infrastructure and training.
- National Horticultural Board (NHB):NHB promotes integrated development in horticulture, assists in development of post harvest management infrastructure, promotes production and processing of fruits and vegetables, strengthening of market information systems and assists in RandD programmes in cultivation and processing. NHB's financial schemes are directed towards commercial horticulture and infrastructure related to post harvest techniques.

Financial assistance from these organizations are in the form of grants, back-ended subsidies, soft loans, refinance etc., with most of the schemes directed to specific sub-sectors of the agri/food processing industry. In addition, some State Governments offer financial schemes to companies in the sector. The key issues with Government schemes are highlighted below:

- Overlap among schemes: Various Government agencies have separate schemes which aim to achieve the same objectives. Thus, an applicant can potentially approach different agencies and obtain assistance from more than one agency. Some of the examples of overlapping schemes are as follows:
- Several areas addressed inadequately: There are a host of areas which are not addressed by these schemes such as financing of freezer cabinets for retail outlets, financing of bulkcoolers for milk, financing of alternate markets/mandis, financing of vending machines for tea/coffee/beverages etc.
- Creation of excess capacity: There are instances where assistance has been provided to set up cold storages in locations which have excess storage capacity. This has lead to an unviable situation for all cold storage operators in the region. It is essential for the nodal

agencies to achieve development of the food processing sector in a sustainable manner.

- Inadequate financial assistance^ most instances, there is a cap on total financial assistance provided at INR 5-5.5 mn. This leads to fragmentation in capacity creation, and often, a situation of overcapacity. At the same time, it does not allow players to scale up their operations.
- Inadequate Monitoring of progress of schemes: It is important to track progress of the projects funded by the Government. Periodic appraisal of the progress will ensure that funds are appropriately utilized. This will also enable assessment of success and lacunae of these schemes and record the multiplier effect of the financial support provided.
- Most of the schemes are back-ended which leads to a funds crunch during implementation.
- Institutions providing subsidies do not undertake their own assessment and rely on the appraisal report of the bank which provides loans to the project.
- There are significant time lags from the date of application for financial assistance, to release of funds, and may affect the project schedule, and lead to cost overruns
- These schemes do not address working capital requirements, which due to the nature of the industry are significant and often higher in terms of quantum than capital investment requirements
- Applicants claim that they are often unaware that their applications have been rejected on account of lack of communication from the funding institution. Further, applicants are unaware of the reasons for rejection.

FINANCING - FOOD PROCESSING SECTOR

The food processing sector comprises a large number of small and medium sized companies, a significant proportion of which have stand alone operations, with no control over the raw material base and reliant on other organizations to undertake marketing/further processing of their products. Banks and financial Institutions adopt the same risk models relevant to the manufacturing sector, for assessing food processing companies. Interest charges for food processing companies are high, on account of the high risk perception associated with the nature of their operations.

KEY ISSUES IN FINANCING: FOOD PROCESSORS' PERSPECTIVE

With raw material availability often being seasonal, and/or concentration in demand, inventory holdings are high and thus the working capital finance issued through normal Maximum Permissible Bank Finance method does not address the funding requirements adequately. The rate of interest for working capital assistance is high. Most food processing companies are not able to access adequate working capital at reasonable rates. This affects the raw material procurement and capacity utilization through out the year. Unavailability of large working capital facilities, or facilities without a high seasonal drawal, restricts players in purchasing agri produce in large volumes, when prices are favourable.

KEY ISSUES IN FINANCING FOOD PROCESSORS: BANKER'S PERSPECTIVE

- Lack of reliable information on demand-supply, price trends, raw material supplies etc is a significant constraint for banks, impacting credit assessment and monitoring
- High operational/transaction cost in servicing small companies
- Lack of backward linkages for assured access to raw material
- Lack of forward linkages of stand-alone food processing units for marketing/distribution

Owing to high cost of market development, the profitability of food processing companies is under pressure in the initial years of their operation. This impacts the risk rating and leads to high cost of borrowing. Even temporary adversities in market conditions can lead to defaults and eventually the loan can become a non performing asset for banks.

CENTRE FOR ADVANCEMENT OF SUSTAINABLE AGRICULTURE

India's record of progress in agriculture over the past four decades has been quite impressive. The agriculture sector has been successful in keeping pace with rising demand for food. The contribution of increased land area under agricultural production has declined over time and increases in production in the past two decades have been almost entirely due to increased productivity. Contribution of agricultural growth to overall progress has been widespread. Increased productivity has helped to feed the poor, enhanced farm income and provided opportunities for both direct and indirect employment. The success of India's agriculture is attributed to a series of steps that led to availability of farm technologies which brought about dramatic increases in productivity in 70s and 80s often described as the Green Revolution era.

The major sources of agricultural growth during this period were the spread of modern crop varieties, intensification of input use and investments leading to expansion in the irrigated area. In areas where 'Green Revolution' technologies had major impact, growth has now slowed. New technologies are needed to push out yield frontiers, utilize inputs more efficiently and diversify to more sustainable and higher value cropping patterns.

At the same time there is urgency to better exploit potential of rainfed and other less endowed areas if we are to meet targets of agricultural growth and poverty alleviation. Given the wide range of agroecological setting and producers, Indian agriculture is faced with a great diversity of needs, opportunities and prospects. Future growth needs to be more rapid, more widely distributed and better targeted. These challenges have profound implications for the way farmers' problems are conceived, researched and transferred to the farmers. On the one hand agricultural research will increasingly be required to address location specific problems facing the communities on the other the systems will have to position themselves in an increasingly competitive environment to generate and adopt cutting edge technologies to bear upon the solutions facing a vast majority of resource poor farmers.

In the past agriculture has played and will continue to play a dominant role in the growth of Indian economy in the foreseeable future. It represents the largest sector producing around 28 percent of the GDP, is the largest employer providing more than 60 percent of the jobs and is the prime arbiter of living standards for seventy percent of India's population living in the rural areas. These factors together with a strong determination to achieve self-sufficiency in food grains production have ensured a high priority for agriculture sector in the successive development plans of the country.

An important facet of progress in agriculture is its success in eradication of its critical dependence on imported foodgrains. In the 1950's nearly 5 percent of the total foodgrains available in the country were imported. This dependence worsened during the 1960's when two severe drought years led to a sharp increase in import of foodgrains. During 1966 India had to import more than 10 million tonnes of foodgrains as against a domestic production of 72 million tonnes. In the following year again, nearly twelve million tonnes had to be imported. On the average well over seven percent of the total availability of foodgrains during the 1960s had to be imported.

Indian agriculture has progressed a long way from an era of frequent droughts and vulnerability to food shortages to becoming a significant exporter of agricultural commodities. This has been possible due to persistent efforts at harnessing the potential of land and water resources for agricultural purposes. Indian agriculture, which grew at the rate of about 1 percent per annum during the fifty years before independence, has grown at the rate of about 3 percent per annum in the post independence era.

AGRICULTURE – SUB-SECTORS

Indian agriculture broadly consists of four sub-sectors. Agriculture proper including all food-crops oilseeds, fibre, plantation crops, fruits and vegetables is the largest accounting for nearly 70 percent of the agriculture sector as a whole. The rapid growth in this sub-sector through exploitation of wastelands and fallows, spread of irrigation and adoption of production enhancing technologies was critical in transforming India from a country vulnerable to food shortages to one of exportable surplus. Although this sub-sector has made impressive progress its share in the sector as a whole has declined from 78 percent in 1960-61 to less than 70 percent by early 90s.

Correspondingly the share of livestock sector has increased considerably. The livestock industry has grown from ₹ 15 billion in early 1960s to ₹ 100 billion by 1980-81 and ₹. 672 billion by 1993-94. In nominal terms the sector grew at almost 15 percent per annum during 1980s. Milk production, which was almost stagnant for two decades ending 1970, grew by over 5 percent per annum in the 80s. Similarly, production of eggs increased at the rate of about 6.5 percent during the same period. As a result the share of livestock increased from about 17 percent till early 80s to 25 percent by 1993-94.

Though it plays relatively a minor role within the sector as a whole, fishing sub-sector activities have been on the rise. The sub-sector has grown from only ₹. 3 billion in 1970-71 to nearly ₹. 90 billion in 1993-94. The growth was particularly rapid in 70s and 80s. Value added increased at over 5 percent per annum during this period.

In real terms forestry and logging activities have been on the decline since mid seventies. As of 1993-94, the size of the industry in terms of value of output was 103 billion.

Over the past three decades, the country has successfully transformed itself from a food deficit economy to one which is essentially self sufficient in availability of foodgrains and other essential commodities, albeit only at the prevailing level of effective demand. Annual aggregate foodgrains production, which averaged about 82 million tonnes in 1960-61 increased to 123.7 and 172.5 million tonnes for the trienniums ending 1980-81 and 1990-91 respectively. Current (1998-99) production level is 195 million tonnes and the country has been able to accumulate substantial, (35 million tonnes) stocks of foodgrains to cope up with any sudden difficulties arising from drought or a similar situation in any part of the country.

Increased outputs, have been achieved chiefly by adopting, since mid sixties, a strategy aimed at increasing foodgrains production by concentrating public sector efforts and resources in regions with a high potential for quick and substantial productivity gains through increased cropping intensity and average yields. These were the areas favoured by agroclimatic resource conditions and where irrigation facilities already existed or could be developed relatively rapidly. The main elements of this strategy were: (i) expansion of irrigation coverage, (ii) increased provision and utilization of key inputs – mainly high yielding varieties (HYVs) of crops, mainly of wheat and rice and chemical fertilizers and plant protection chemicals, (iii) expansion and improvement of institutional support services such as research and extension and (iv) price policies favourable to producers of major foodgrains. The success of this strategy was made possible by development and availability of replicable production technology packages, so called 'Green Revolution' technologies. Irrigation facilitated double cropping and widespread adoption of HYVs. The HYVs performed particularly well under irrigated conditions, were highly responsive to fertilizers and their short duration permitted increases in cropping intensities.

Irrigation development was the cornerstone of the strategy. Undivided India was amongst the largest irrigated areas in the world. With partition nearly one-third of the irrigated area went to Pakistan. At the time of independence the net irrigated area was 20.9 million ha (gross irrigated area 22.6 million ha). Recognizing large-scale development of irrigation facilities as critical to rapid agricultural growth, the country has spent about ₹. 45,000 crores on irrigation development in the first four decades after independence. During the period 1950-51 to 1965-66 development of irrigation through government canals grew from 7.2 million ha to 9.8 million ha – a growth rate of 2.1 percent per annum.

During 1970s this pace dropped slightly to 1.9 percent. In 1980s the rate of increase dropped significantly to 1.1 percent per annum. The growth of tube-well irrigation, however, increased rapidly from 4.5 million ha in 1970-71 to 9.5 million ha in 1980-81 and then to 14.3 million ha by 1990-91. The net irrigated area increased from 31 million ha in 1970-71 to 53.5 million ha in 1995-96 which corresponds to 22 percent of the net sown area in 1970-71 and 37.63 percent in 1995-96. With improvements in irrigation efficiency the gross irrigated areas has increased to 71.51 million ha. The percentage of gross cropped area service by irrigation increased from 18.3 percent in 1960-61 to 23.0 percent in 1970-71 and to over 38 percent at present.

Fertilizers have constituted yet another key input in addition to expanded irrigation and spread of HYVs in achieving goals of high production and productivity. India currently occupies third position in the world, after China and USA, in terms of fertilizer production and consumption. Consumption of fertilizers has increased from 1.54 million tonnes in 1967-68, representing the pre green revolution era, to 17.31 million tonnes currently (1997-98). The average per hectare use of fertilizers currently around 85 kg per hectares is the lowest among several Asian countries.

However, rice and wheat account for a major fraction, around 65 percent of the total fertilizer consumed in the country, with very little fertilizers going to the rainfed areas. According to some current projections, fertilizer's use will need to increase to 30-35 million tonnes to meet the foodgrains need of 2020. The demand for nutrients will stretch by almost another 15 million tonnes if requirements for horticulture, vegetables and plantation and commercial crops are included. At present domestic production of N and P fertilizers (13.42 million tonnes) falls short of consumption by over 20 percent. In addition the entire requirement of K fertilizer is imported.

Most agricultural development programmes initiated in 1960s were concentrated in regions of high potential. Thus five states, Punjab, Haryana, Uttar Pradesh, Andhra Pradesh and Tamil Nadu account for 50 percent of the country's net irrigated and 53 percent of the gross irrigated area. The combination of expanding irrigation coverage and widespread adoption of short duration HYVs led to significant increases in cropping intensities. Acreage cropped more than once per year increased from 13 million ha in 1950-51 to about 44 million ha at present. Average cropping intensity for the country as a whole rose from 115 percent in 1960-61 to 131 in 1993-94. By 1993-94 cropping intensity has risen to 187 percent in Punjab, 167 percent in Haryana and 142 percent in Uttar Pradesh.

Production of foodgrains other than rice and wheat did not increase significantly and in the eastern region even the yield of rice did not increase. Agricultural production and income rose substantially in the north-western states of Punjab, Haryana, Western Uttar Pradesh, parts of Rajasthan, Tamil Nadu and Andhra Pradesh. By contrast productivity and output growth have been modest in eastern and central India and in deccan plateau. Progress was particularly slow in rainfed areas, which account for over 60 percent of the cropped area and where a great majority of rural poor are concentrated.

An important impact of the strategies pursued in the 'Green Revolution' period has been intensification of regional disparities and imbalances in agricultural development and food availability and hence levels of food security. For the country as a whole while per capita availability of cereals has increased substantially, that of pulses has decreased significantly. In summary an annual increase in foodgrains production of 3.22 percent during fifties was mainly because of expansion in area. Sixties recorded a low annual growth rate of 1.72 percent necessitating large-scale imports of foodgrains. Annual growth of 2.08 percent was recorded during seventies. This decade was the turning point in India's foodgrains economy leading to self-sufficiency through significant productivity increase first in wheat and later in rice in the eighties. An annual growth of 3.5 percent in foodgrains in eighties was the hallmark of green revolution that enabled India to become self-sufficient and even a marginal exporter. The pace of growth slowed in nineties barely making or even slower than the population growth rate. This is a matter of concern.

HORTICULTURE

The diversity of physiographic, climate and soil characteristics enables India to grow a large variety of horticultural crops – fruits, vegetables, flowers, spices, aromatic and medicinal plants, plantation crops etc. India is the largest producer of fruits in the world and second largest producer of vegetables. The area under fruits estimated at 1.45 million ha in 1970-71 grew to 2.8 million ha in 1991-92 and then more rapidly to 5 to 6 million ha by 1994-95. This sector is likely to grow rapidly in the future both on account of internal demands and export opportunities.

ANIMAL HUSBANDRY AND FISHERIES

Animal husbandry and dairying sub-sector plays an important role in overall economy and in social development. The contribution of the sub-sector is estimated to be about 25 percent of the total value of output of agricultural sector. The sector also plays a significant role in supplementing family incomes and generating employment in the rural sector particularly among the landless, small and marginal farmers and women besides providing nutritious food. Production of suitable cross breeds and their wider adoptions has contributed to increasing country's milk production, which has now reached 75 million tonnes annually. Similarly, genetic improvement and better management practices have markedly pushed up production of poultry and eggs.

Through the overall contribution of fisheries sub-sector is small, development of prolific and fast growing forms of several common fish species coupled with breakthrough in breeding under captivity, fish seed production and multi-layer fish culture has resulted in registering a very high annual growth of 11 percent in aquaculture production during the past decade.

SUSTAINABILITY CONCERNS

Several indicators highlight increasing concerns of sustainability in areas which have largely contributed to increased production in the 'Green Revolution' era. Adoption of high yielding cultivators is virtually complete. Almost entire wheat and rice crops in the states of Punjab, Haryana and Western Uttar Pradesh are irrigated. In the higher production regions yields are plateauing and most traditional sources of productivity growth having been exhausted future gains in production have to come from elsewhere.

At farmers' level concerns are being expressed in several ways. Many farmers believe that the input levels have to be continuously increased in order to maintain high yields. In sixties and seventies most farmers used only nitrogenous and phosphate fertilizers to achieve high yields. Due to widespread deficiencies of several secondary and micronutrients, most farmers now have to apply higher doses and a greater variety of fertilizers to maintain crop yields. Results from many long term studies on rice-wheat cropping system show a declining yield trend when input levels were kept constant – thus the growth rate of system productivity has been declining relative to growth rate of nutrients use. Lowering of groundwater tables due to intensive rice-wheat system in many areas is resulting in increased costs of lifting water in the intensively cultivated high production areas, diseases and pest problems are turning more serious than ever before and pose both short and long large problems. It is reported that some weeds have developed resistance to the commonly used herbicides. What this implies is that the farmers are applying increasing amount of herbicide incurring increasing cost without the benefit of effective control. Pesticide residues entering the food chain and overall safety in use of pesticides continue to be serious problems.

Other emerging problems threatening sustainability of intensive cropping system *e.g.* rice-wheat include loss in biodiversity related issues. Large areas planted to a single/few varieties of a crop is a potential cause of concern. As the diversity is reduced natural processes that control and affect habitat quality and genetic expression weaken and for this reason internal and natural control mechanisms must be replaced by more externally applied artificial controls in the form of management and inputs which in due course lead the system towards unsustainability.

Groundwater is the major source of meeting the irrigation needs of irrigated agriculture. Currently about half the area under irrigation in the country is irrigated from groundwater sources. Large-scale groundwater development has led to fall in the water table in many areas. Over pumping is leading to declining water table levels and failure of tube-wells. Pumping costs are increasing, as is the energy consumption. In the coastal areas this has led to ingress of sea water, with serious environmental implications.

Changes in water quality are adversely affecting agriculture and vice-versa. Inefficient and/or over use of fertilizers and pesticides in agriculture and untreated disposal of industrial and urban wastes are leading to increasing contamination by such elements as lead, zinc, copper, chromium, cadmium particularly in areas having high industrial activity *e.g.* in districts of Ludhiana, Faridabad, Kanpur, Varanasi etc.

An increase in the content of arsenic has been reported in several of the districts of West Bengal. This is attributed amongst other causes to the lowering of groundwater table due to excessive groundwater withdrawal and is leading to serious and widespread toxicity problems adversely affect the health of hundreds of thousands people of the region.

THE INTERNATIONAL COMMISSION ON PEACE AND FOOD

The Indian nation is on the move. The people are astir. Agriculture is blooming. Industry is booming. The legacy of self-deprecation and self-doubt inherited from the colonial period have given way to a new-found dynamism, confidence and sense of self-determination. No longer harping critically about its past, the country has come to recognize its commendable achievements and become aware of the tremendous opportunities which the future holds for even more rapid progress. This change in attitude and this emerging consciousness of opportunity are the most essential ingredients for accelerating development.

STRATEGY TO COMPLEMENT THE ECONOMIC LIBERALIZATION

The international community has noted with some amazement India's impressive gains over the last five years in recovering from the severe economic problems brought on by high rates of inflation, high budget deficits, and falling foreign exchange reserves. Today the country enjoys a very low inflation rate, comfortable reserves of hard currency and an industrial growth rate approaching 12per cent. Exports are growing at a phenomenal rate of 24per cent. Employment generation has risen from an average of 4.8 million new jobs per annum in the 1980s and a low of 3 million just five years ago to a record 7.2 million last year.

These gains are remarkable and beyond the expectations of even the most far-sighted analysts. However, even if the most optimistic targets are achieved, it will not ensure that rising incomes and employment opportunities will be available to all who presently live in poverty.

The problem with the economic reform package is not that it is stimulating the rapid growth of some sectors, especially industry and exports, but that it is not doing enough for the 70per cent of the population residing in rural areas, of which 65per cent are engaged in agriculture. India needs a new strategy to complement the economic reforms, a strategy designed to take full advantage of the country's agro-climatic advantages, huge tracks of cultivable land, and large internal market to stimulate a rapid development of the rural economy.

Agriculture as an Engine for Industrialization: Rising productivity and profitability from agriculture was a primary cause of the Industrial Revolution in England during the early 19th Century. It raised the purchasing power of the rural population, stimulated greater demand for manufactured goods and led to the growth of industries. In America too, rising productivity in agriculture was a strong spur to the development of the manufacturing sector at the turn of the century.

The rapid development of South Korea, Taiwan, and Malaysia provides ample evidence that an agriculture-driven strategy can act as a powerful engine for raising rural incomes, industrialization and employment generation. More recently, commercial agriculture has been a powerful engine for economic development in Thailand, where 70per cent of the population is still employed in the farm sector.

India's Competitive Advantage in Agriculture: India possesses four outstanding competitive advantages in agriculture comparable to those of any other country in the world. First, it has regions which are climatically favourable for cultivation of every commercially-important plant species grown in other parts of the world – ranging from temperate orchard crops such as almonds and apples to tropical mangoes and pineapple. Second, the country already possesses the largest acreage of irrigated land in the world with 40per cent of the potential still to be tapped. Third, the gap between present productivity and proven technological potential is very large for most crops; yet even so, the country is already among the world's top three producers of tea, cotton, sugar, food grains, groundnut, coffee, eggs and milk. Fourth, the country has an abundance of available skilled, educated, technical and scientific manpower.

From Minimum Needs to Maximum Potentials: During the colonial period India suffered from food deficits and severe famines. Thus, in the early days of India's Independence, it was natural for the Government to focus its attention on production of sufficient food grains to meet the minimum dietary needs of a rapidly expanding population.

As a result of the Green Revolution, India defied the dire predictions of international experts by doubling its food grain production within a decade and achieving self-sufficiency in cereal crops. Since then, the agricultural sector has ceased to receive the attention it deserves. Combating the threat of famine is no longer a pressing priority, but that does not mean agriculture has fulfilled its mission.

It is time to stop thinking of agriculture simply as a means to provide for the minimum food needs of an expanding population. We must come to look upon agriculture as a business and do everything possible to maximize productivity, incomes and employment in this sector. This is as important for the individual farmers trying to maximizing the living standards of their families with limited land and financial resources as it is for companies seeking profitable investments in the agri-business sector.

Rising farm incomes will generate tremendous demand for consumer goods, equipment, processing and distribution industries stimulating rapid growth in industry and rising job opportunities in both the rural and urban sectors. Here lies a strategy that can bring prosperity to the Indian people.

Creating Rural Jobs as a Counter to Urban Migration: Agriculture is the key to generating full employment in the country. This sector not only provides on-farm employment but also generates employment opportunities in processing industries, manufacturing, food distribution and the transport sector. Although less than 3per cent of the US population is currently engaged in farming, one out of every three jobs in America is related to the production, processing, distribution or export of agricultural produce.

Wherever agriculture has become modernized in India, labour has become scarce. Progressive agricultural centers such as Coimbatore, Pune and Punjab have also become centers of industrialization, where many farmers shy away from labour intensive crops because workers are finding higher paid employment opportunities in industry.

Full development of India's potential for commercial agriculture, agro-industry, agri-business and agro-exports can eradicate poverty and unemployment within the decade. A study undertaken for the International Commission on Peace and

Food concluded that full development of the India's agricultural potential can lead to the creation of more than 100 million new jobs. The study indicates that the cost of creating employment through this approach is by far the most cost-effective and affordable strategy. In fact, it can be accomplished with the available resources and without dependence on external resources by a reprioritization of plan investments.

A detailed micro-level study of Pune District by the Agricultural Finance Corporation has confirmed this potential. AFC's report identifies a wide range of commercially viable opportunities for both private farmers and the corporate sector, and estimated that 750,000 jobs can be created in this single district through an agriculture-led strategy. Similar studies need to be conducted throughout the country to fully document the potential at the local level.

According to one estimate, *for every ton of additional foodgrains produced in India, one new job is created in the economy.* Rising productivity in agriculture can stimulate the growth of agro-industries, food processing and distribution, and demand for new industrial plants and machinery. It will also increase demand for consumer good, household appliances and tourism in the rural sector, creating a boom in related sectors.

GROWING DOMESTIC DEMAND

Achieving nutritional security for the Indian masses will require a significant increase in daily intake of calories, protein, vitamins and minerals. The present nutritional gap represents a vast pent-up demand for more and better quality foods while can be translated into commercial opportunities for India's farmers. Indians consume an average of 40 grams per day of horticulture products compared to a normal nutritional demand of 90 grams. As incomes rise, the domestic market for horticulture products is projected to increase by 60per cent over the next six years. Overall, there is scope for placing an additional 1 million acres under horticulture crops. This will generate demand for 100 new commercial hybrid seed production units in the country.

Sugar consumption in India has tripled over the last three decades and now exceeds 13 kg. per person per year. Yet even current levels of consumption remain very low compared with those of other developing countries: 21 kg. in Kenya, 31 kg. in Argentina, 33 kg. in Egypt, 44 kg. in Brazil and 45 kg. in Mexico. Over the next decade, rising incomes are projected to increase India's domestic consumption to 25 kg. per capita. This will create demand for at least 300 new sugar mills. Failure to anticipate surging demand could lead to massive sugar imports.

TREMENDOUS EXPORT OPPORTUNITIES

India has the potential to become a global leader in agriculture. Already agriculture exports, including textiles, have risen from ₹ 10,000 crores to ₹

40,000 crores over the past five years. Grape and mango exports to Western Europe are rapidly increasing.

Floriculture is a $40 billion global industry that is projected to reach $70 billion by year 2000, of which international trade in cut flowers accounts for $6 billion. India's exports of cut flower have risen from $2 million to $10 million annually over the past five years, but the potential is 100 fold greater. New floriculture projects are already springing up around the country.

Exports of processed fruits and vegetable, cotton textiles, sugar, and fish have vast potential. Processing can multiply the export value of farm produce by 50 to 500 times and open up vast international markets.

India currently processes less than 2per cent of its agricultural produce compared with 30per cent in Brazil, 70per cent in USA and 82per cent in Malaysia.

India's mushroom exports are mushrooming. Over the past three years not less than nine export-oriented mushroom projects with an investment of more than ₹.130 crores have been established in joint venture with foreign companies.

Despite being the second largest producer of silk in the world, India's share of world silk trade is less than 5per cent. Improving the quality of domestic silk production offers a lucrative opportunity for exports.

CLOSING THE PRODUCTIVITY GAP

The potential for raising agricultural productivity is enormous. India ranks at or near bottom of the list of countries in the world in terms of productivity per acre for almost all major crops.

With 60per cent more arable land, India produces less than half the quantity of foodgrains grown by China. Brazilian yields of black pepper are six times higher than India utilizing varieties original imported from India. In rice, Indian yields are one third the level in North Korea. Maize yields are one-sixth the level of Chile.

Wheat yields are one-third Ireland's average. Soyabean yields are one-third the level achieved in South Africa. Productivity in pulses is one-tenth the level of France. In groundnut Indian yields are one-sixth Israel's average.

India agriculture suffers from low productivity of its soil and water resources. Raising productivity means increasing profitability. Indian farmers have vast scope for generating higher incomes in agriculture.

The average yield of tomato in India is 12 tons per acre versus 34 tons in the USA, but yields as high as 38 tons have been achieved by commercial farmers in India generating net profit in excess of ₹ 50,000 per acre.

Average net incomes ranging between ₹ 40,000 and ₹ 1,00,000 per acre are now being achieved by some modern farmers employing advanced cultivation practices on a range of vegetable, flower and fruit crops.

OPTIMIZING INDIA'S SOIL AND WATER POTENTIAL

A shift in perspective is needed from viewing agriculture merely as an occupation and source of food, to recognizing that it is a business. Regardless of whether the farming is done by a small cultivator with one or two acres or a large corporation, the aim is the same – to generate the highest yields and income per unit of land, water and capital employed.

From a commercial perspective, India is generating less income for its farmers per unit of available land and water. This low water and soil productivity can be overcome by adopting proven modern technologies for soil restoration and water conservation.

Thus far Indian farmers and scientists have focused on raising productivity through the application of marco-nutrients, nitrogen, phosphorus and potassium, while largely ignoring the crucial role of micro-nutrients in bringing forth the full genetic potential of plant materials. According to demonstrations recently carried out in several regions of India under the direction of Dr. C. Lakshmanan, a California-based *Cabbage harvest by Bihar tribals* agricultural consultant, it has been shown that the timely application of missing micro nutrients based on scientific soil tests can double the yields of most crops in the very first year and double net incomes too. These results have been achieved while significantly reducing the input of macro-nutrients. This approach has enabled tribal farmers in Bihar to dramatically improve maize productivity and to earn an average income of ₹ 75,000 per acre from cabbage.

Despite the widespread belief that shortage of water is a critical constraint in Indian agriculture, the fact is that water is often being wastefully used due to inefficient irrigation technology. Indian farmers visiting American farms were surprised to learn that their US counterparts stop watering tomato 45 days before harvest compared to two days here under desert-like climatic conditions as severe as those in India, yet the US farmers obtain yields up to six times the Indian average. In many cases excess of water is washing away precious nutrients, reducing soil fertility and inhibiting plant growth. Application of water saving technologies can raise the productivity of water in Indian agriculture by two, three or four times its present level, resulting in higher yields, greater production, higher incomes and more jobs.

CREATING RURAL ENTREPRENEURS

With the rights technology and management practices, agriculture offers lucrative for entrepreneurs. A ten acre farm growing a combination of cash crops such as sugar, cotton, fruits, vegetables and flowers can earn ₹. 5 lakhs or more per annum for a rural family.

Although it is widely believed that the problem of educated unemployment cannot be solved, a shortage of agricultural graduates is actually developing due to the rapid development of commercial agriculture. A single agri-business enterprise has recruited more than 250 agricultural graduates within the past two years. When the commercial potentials of agriculture are fully recognized, students will flock to agricultural colleges and universities as a course of preference and many agricultural graduates who come from rural families will return to the land to become entrepreneurial commercial farmers, rather than migrating to urban areas in search for employment. *Professional farmers should come to acquire the social status and prestige once accorded to government employment and now granted to industry.*

RURAL AQUACULTURE ESTATES

Fresh water fish and prawn culture can be a highly remunerative undertaking for rural farmers, provided that they have access to appropriate technology, feeds, processing and marketing facilities. One approach is for Government to establish rural aquaculture estates and lease out small production ponds to farmers and landless labour. Each estate could consist of 20 to 25 acres of ponds, each pond.25 acres in size and equipped with machinery for aeration and drainage. Each estate should have a centralized source of pure water, power, feed and processing plant, cold storage and marketing facilities. Utilizing fast growing varieties, each pond can generate a minimum of ₹ 50,000 per acre net profit per annum.

New models and innovative approaches are needed to bring small producers together to form viable rural enterprises. One option is for groups of farmers to constitute their own firms for joint production, processing and marketing. Integrated horticulture corporations owned by the growers can coordinate production, processing and marketing of a range of fruits and vegetables grown over a 1000 acre extent. Integrated sericulture projects can be established in which all the essential operations from mulberry cultivation to spinning of silk yarn and weaving of silk fabric can be brought together in a small cluster of villages, minimizing the need for middlemen and maximizing profits to the primary producers.

ROLE FOR THE CORPORATE SECTOR

But efforts to organize the farm community should not blind us to the very constructive and valuable role which the corporate sector has played and can continue to play in generating rural prosperity. As a result of the exploitation of the country by foreign firms during the colonial period, it came to be widely accepted that the interests of the business community and the interests of the Indian masses were irreconcilably opposed to one another. Despite the development of a large domestic entrepreneurial class and a ten-fold growth of private and public limited companies during the past quarter century, this out-dated thinking continues to obscure the opportunities for mutually beneficial cooperation between the formal and informal sectors of the economy. The network of producer dairy cooperatives linked into a national milk grid by the National Dairy Development Board and the linkage between small cane growers and large sugar mills for processing and marketing sugar are examples of successful models that can be extended to other crops. Similar models of cooperation can and should be established for promoting horticulture crops, poultry, and freshwater aquaculture.

In some cases the country will benefit by encouraging private sector firms to become primary producers as well. India has over 100 million hectares of uncultivated and degraded wastelands which is not generating any benefit either to the rural population or the country as a whole. Large tracks of this land can be converted into productive cultivable land by an infusion of capital and sophisticated technology to tap deep aquifers, install drip irrigation facilities and in some cases green houses. The cost and technical input required to develop these lands may be far beyond the means of small farmers in the area, but can be undertaken by agri-business corporations. Already some 200 companies have taken up commercial agriculture, restoring productivity to degraded wastelands and generating both on-farm and off-farm employment opportunities for the rural population. One firm has placed 18,000 acres of uncultivated land under intensive cultivation of horticulture crops

The corporate sector can also play a role in stemming and reversing the degradation of India's forests. Out of 67 million hectares classified as forest, only about 60per cent presently has good tree cover. Remote sensing data indicate the covered area is declining at the rate of 1.5 million hectares annually. The task of reversing this degradation is beyond the means of government to accomplish on its own.

Appropriate policies need to be formulated to encourage the private sector to invest in planting the barren areas and farming portions of the tree crops which they raise under contractual agreements with the forest authorities, according to practices commonly by many highly industrialized nations. Such arrangements will generate job opportunities and enhance supplies of much needed wood products.

INDIA'S UNIQUE OPPORTUNITY

Never before has India been poised for such rapid economic progress. Never before have the opportunities been so great or the country so ready. Now what is needed is to generate a widespread awareness of the potentials at all levels of the society – among political leaders, administrators, farmers and the corporate sector.

It took the leaders of India's Freedom Movement more than half a century to achieve Independence largely because it took that long to convince the Indian elite and masses that such a goal was achievable.

It has taken another half century for the Indian nation to cast off the oppressive legacy of colonialism and to rediscover its own inherent dynamism and resourceful.

It need not take another half century to convert this rich capacity into material accomplishments. What is needed now is a national Prosperity Movement to realize the goal of Indian economic accomplishment. If pursued with confidence and determination, India can achieve widespread prosperity within a decade.

ACTION PLAN FOR PROSPERITY 2000

- Establish commercial farming schools on leased lands in every block to demonstrate cultivation of highly profitable cash crops and train young farmers in advanced methods to raise productivity.
- Establish 1000 integrated horticulture estates covering one million hectares of irrigated land, for private farmers to cultivate high profit vegetable and fruit crops linked to professional processing and marketing by private or public sector agencies.
- Establish 500 integrated sericulture projects in which all the operations from mulberry cultivation to silk spinning and weaving are carried out scientifically within a clusters of villages and the products are professionally marketed and exported.
- Establish 2500 intensive aquaculture estates, each of 50 acres, consisting of quarter acre intensive production ponds leased out to small farmers and landless workers with centralized technical support, feed plant, processing, storage and marketing facilities.
- Establish scientifically run soil labs in every district to test soils for micro-nutrients and prescribe measures to restore soil fertility and double crop productivity while reducing inputs of macro-nutrient chemical fertilizers.
- Encourage the private sector to acquire or lease degraded, uncultivable waste lands and to utilize advanced technologies to reclaim 10 million acres for intensive horticulture and farm forestry.
- Revamp the curriculum of agricultural colleges and universities to impart practical skills in commercial farming and to encourage graduates to take up scientific farming and agri-business ventures.
- Widely publicize achievements in the agri-business sector to generate awareness of the enormous potential for the country.

AGRICULTURE IN THE CHANGING GLOBAL SCENARIO

Steady globalization of trade has profound implications for future agricultural development. The diversity of India's agro-ecological setting, high bio-diversity and relatively low cost of labour provide potential for agricultural competitiveness in a globalized economy. It is expected that with increasing globalization of markets over the years there will be demands for agricultural intensification.

This will also be favoured because of greater backward and forward linkages between agriculture and food industry. Therefore, increase in production and productivity are bound to be strategically important to economy. Intensification will not only favour alleviation of rural poverty but will also improve resource conservation particularly in the small farming sector where farmers can be encouraged to take up organized production of high value crops such as fruits, specialty vegetables, flowers medicinal and aromatic herbs etc. Stronger demands for crops of the small farmers' will not only improve incomes and welfare but will also make investments in technology and resource conservation more attractive.The General Agreement on Tariff and Trade (GATT) and liberalization of global trade is bound to have impact on future land use and production pattern. Understanding the local, national and international environment under which agricultural production is taking shape will be crucial in developing our own strategies.

EXTENSION STRATEGIES

Since early fifties a number of public by funded agricultural development programmes have been sponsored. These have included programmes like the National Extension Service (NES) Blocks in 1953, the Intensive Agricultural District Programme (IADP) in 1961-62, the Intensive Agricultural Area Programme (IAAP) 1964-65, the High Yielding Variety (HYV) programme 1966-67 and the Small and Marginal Farmers' Development Programmes (SMFDP) in 1969-70.

Though these programmes had a perceptible impact the efforts did not get replicated over different areas and categories of farmers. In mid seventies based on pilot level project in Rajasthan Canal and Chambal command area a 'Training and Visit' (TandV) system of extension was promoted in different states. Extension efforts of the Indian Council of Agricultural Research through its research Institutes and the State Agricultural University were largely limited to demonstration of new technologies through such programmes as National Demonstration Project, Operational Research Project, the Lab to Land Programme and the Krishi Vigyan Kendras. However, there appears much to be desired in the way that extension programmes are conceived and implemented.

At present extension programmes are implemented in largely a top-down fashion leaving little scope for localized planning and action. Farmers are almost passive receivers and their involvement in the process of technology generation and adoption is almost absent. Extension services, at present, are almost exclusively in the public sector domain and there is no effort or institutional support for other operators *e.g.* the NGOs, the corporate bodies etc. Extension programmes sponsored by the government operate largely in isolation and there appears a strong need to view the extension programmes as an integral part of the research and development process.

The challenges facing agricultural development call for fundamental changes in our approach to technology transfer/extension programmes. Changes are necessary in the context of changing economic environment following policy adjustments in relation to privatization, deregulation and globalization calling for greater efficiency and effectiveness of the extension system. More importantly there is need for

- Greater emphasis on providing producers with knowledge and understanding needed to overcome the problems or to exploit opportunities of their own specific production systems. Correspondingly there will be a need to de-emphasize 'package of practices' or the blanket recommendations, top down approach followed thus far.
- Shift in the focus of public extension systems from promoting inputs use to one on sustainable management of resources and improvements in the production system as a whole.
- Closer interaction between farmers, extension scientists and production system researchers in diagnosing problems and identifying location specific recommendations emphasizing participation and education rather than being prescriptive.
- Widening the range of extension delivering agencies. While the publicly operated extension systems will continue to be important, there will appear a greater role for NGOs, farmers' associations and corporate sectors in particular situations. Role of commercial suppliers of seeds, agrochemicals, machinery, vaccines and medicines in providing advisories, as is already being done in a limited way, will need to be encouraged and factored into public system's own priorities.
- Wider and more creative use of mass media in tune with current developments in information technology to get information across to the farming community whose ability to overcome constrains at farm level will increasingly depend on access to reliable and up-to-date information.

TECHNOLOGICAL NEEDS AND FUTURE AGRICULTURE

It is apparent that the tasks of meeting the consumption needs of the projected population are going to be more difficult given the higher productivity base than in 1960s. There is also a growing realization that previous strategies of generating and promoting technologies have contributed to serious and widespread problems of environmental and natural resource degradation. This implies that in future the technologies that are developed and promoted must result not only in increased productivity level but also ensure that the quality of natural resource base is preserved and enhanced. In short, they lead to sustainable improvements in agricultural production.

Productivity gains during the 'Green Revolution' era were largely confined to relatively well endowed areas. Given the wide range of agroecological setting and producers, Indian agriculture is faced with a great diversity of needs, opportunities and prospects. Future growth needs to be more rapid, more widely distributed and better targeted. Responding to these challenges will call for more efficient and sustainable use of increasingly scarce land water and germplasm resources. Technical solutions required to solve problems will be increasingly location-specific and matched to the huge agroecological/climatic diversity. Detailed indigenous knowledge and greater skills in blending modern and traditional technologies to enhance productive efficiency will be more than ever before, key to the farming success and sectoral growth. Most technological solutions will have to be generated and adapted locally to make them compatible with socio-economic conditions of farming community.

New technologies are needed to push the yield frontiers further, utilize inputs more efficiently and diversify to more sustainable and higher value cropping patterns. These are all knowledge intensive technologies that require both a strong research and extension system and skilled farmers but also a reinvigorated interface where the emphasis is on mutual exchange of information bringing advantages to all. At the same time potential of less favoured areas must be better exploited to meet the targets of growth and poverty alleviation.

These challenges have profound implications for products of agricultural research. The way they are transferred to the farmers and indeed the way research is organized and conducted. One thing is, however, clear – the new generation of technologies will have to be much more site specific, based on high quality science and a heightened opportunity for end user participation in the identification of targets. These must be not only aimed at increasing farmers' technical knowledge and understanding of science based agriculture but also taking advantage of opportunities for full integration with indigenous knowledge. It will also need to take on the challenges of incorporating the socio-economic context and role of markets.

With the passage of time and accelerated by macro-economic reforms undertaken in recent years, the Institutional arrangements as well as the mode of functions of bodies responsible for providing technical underpinning to agricultural growth are proving increasingly inadequate. Changes are needed urgently to respond to new demands for agricultural technologies from several directions. Increasing pressure to maintain and enhance the integrity of degrading natural resources, changes in demands and opportunities arising from economic liberalization, unprecedented opportunities arising from advances in biotechnology, information revolution and most importantly the need and urgency to reach the poor and disadvantaged who have been by passed by the green revolution technologies.

Another important implication of increasing globalization relates to the need for greater attention to the quality of produce and products both for the domestic and the foreign markets. This would imply that production must be tuned to actual rapidly changing product demand. Such adaptation to global markets would require state of the art research, which can be achieved only by setting global standards of research, focus on well defined priorities and mechanisms which permit close interaction of farmers with researchers, the private sector and markets.

CHANGING LAND-USE AND FUTURE OF AGRICULTURE

One of the most important consequences of growing pressure on land is the declining trend in the average farm size and the pattern of holdings. According to the latest Agricultural Census in 1970-71 there were 70 million holdings operating 162 million ha. By 1990-91 there were 105 million holdings operating 165 million ha. The average farm size decreased from 2.30 ha in 1970-71 to 1.57 ha in 1990-91.

As of 1990-91 about 78 percent of holdings were small (1.0 to 2.0 ha) and marginal (<1.0 ha). A little more than 20 percent of the farmers were semi-medium (2.0 to 4.0 ha) and medium (4.0 to 10.0 ha).

Large farmers (>10.0 ha) constituted only 1.6 percent of the total holdings. Over the twenty-year period since 1970 the proportion of marginal farmer has increased from 50 to 59 percent and that of large farmer has declined from about 4 to 1.6 percent.

The proportion of total area operated by marginal farmers increased from nine percent in 1970-71 to nearly 15 percent in 1990-91 while the proportion of large farmers declined from about 31 percent to 17 percent in the same period. The size of average holding is very unevenly distributed among the states.

States with relatively large average size of operational holding are a mixed lot – they include states with large tracts of barren lands *e.g.* Rajasthan, Maharashtra and Madhya Pradesh on the one hand and agriculturally advanced state like Punjab on the other. These trends in farm size changes will have a profound effect on the future agricultural development strategies.

Although India's population growth rate has slowed from 2.1 percent in 1980s to 1.8 percent in the 1990s and is expected to slow further in the coming decades, yet the population is projected to reach 1.33 billion by 2020 from the current one billion. The urban share of total population is projected to increase from 26 percent to 35 percent of the total population. Although incidence of poverty is falling, it is estimated that in 93-94 (upto which data is available) 320 million people constituting 36 percent of the population were below the officially defined poverty line.

The nature of the poverty line has been shifting. About 30 years ago 48.4 percent of those living in rural areas were poor and 20 percent of those living in the urban areas were classed as poor. Recent studies show that the number of poor in urban areas have been increasing at relatively higher rate compared to the rural areas. At present those below the poverty line in rural sector constitute 37 percent of the population while in the urban sector the percentage is 32 percent. In the context of poverty alleviation, therefore, emphasis will be required to be placed both on production of food by the poor as well as on the availability of food for the urban poor. It needs to be recognized that a large proportion of the rural poor are located in regions of low potential for food production *e.g.* arid and semi-arid areas, hilly regions, degraded land and forest areas. Widespread hunger and malnutrition are the direct manifestation of poverty and will call for increasing efforts to produce more food at affordable price.

Increasing population and economic growth are changing patterns of land use making potentially unsustainable demands on the country's natural resources.

- Since early fifties the net area sown was expanded rapidly at first but at a diminishing rate since 1970 to reach approximately 142 million ha at present. During 1950s and 1960s areas under agriculture expanded substantially as the fallows were reduced and cultivable wastes were put under the plough. The net area sown increased from 119 million ha in 1950-51 to 133 million ha by 1960-61 and further to 140 million ha by 1970-71. Fallow lands declined from 28 million ha in 1950-51 to 20 million ha by 1970-71. Cultivable wastelands declined from 23 to 17.5 million ha.
- Land use intensity *i.e.* fraction of net sown area to total geographical area increase from 36 percent in 1950-51 to 40.5 percent in 1960-61 and 43 percent by 1970-71 where it has since stabilized.
- Cropping intensity *i.e.* gross sown area as percent of net sown area increased from 111 percent in 1950-51 to 115 percent in 1960-61, 118 percent in 1970—71 and 130 percent by mid 1990s.
- While the contribution of increased area in the growth of agriculture has declined over time, that of productivity has increased. The yield of all crops grew at 1.5 percent per annum between early 1950s and mid 1960s. the pace accelerated to 1.7 percent in the 1970s and then to 3 percent per annum between early 1980s and mid 90s. Unlike the gains in area, which benefited non-foodgrains, the gains in productivity accrued mostly to foodgrains.

- India's forest resources have been dwindling. According to the 'State of Forest Report' (1997) the total forest cover of the country is estimated at 63.34 million ha *i.e.* 19.27 percent of the geographic area of the country. Of these the dense forest (crown density more than forty percent) and open forest (crown density 10 to 40 percent) occupying about 11 and 8 percent of the geographic area respectively and mangroves occupy 0.15 percent of the geographic area. The country has lost about 5482 sq. km. of forest cover since the 1995 assessment. By any estimate the area under forest is far below the national policy goals and many areas nominally under forest are being used for non-forest purposes. Similarly 'uncultivated lands' such as permanent pastures, miscellaneous tree crops, cultivable wastes and fallow is subject to increasing competition from uses other than feeding livestock.
- The growth of livestock population is an important source of competition for land.
- The area sown to fodder crops is not recorded. Information available from other sources provide an estimate ranging from 4 to 5.5 percent of the net sown area and suggest that the area under fodder crops will have to increase to 10 percent or more to support increasing livestock based activity.

The pressure on India's land and water resources is seriously threatening native plant and animal diversity. India has uniquely rich and diverse genetic base. With increasing agriculture and economic development the genetic pool is declining. This decline, if unchecked and poorly managed can have unforeseen and adverse consequences for the sustainability of agriculture of the region.

4

Principles of Food Preservation

INTRODUCTION

After learning about the causes of food spoilage, it should not be very difficult to list the principles of food preservation. Remember, a good method of food preservation is one that slows down or prevents altogether the action of the agents of spoilage. Also, during the process of food preservation, the food should not be damaged. In order to achieve this, certain basic methods were applied using the knowledge gained form observation of the effects of natural conditions on different types of foods. For example in earlier days, in very cold weather condition, ice was used to preserve foods. Thus, very low temperature became an efficient method for preventing food spoilage. Let us now list the principles of food preservation:

- *Removal of Micro-organisms or Inactivating Them*: This is done by removing air, water (moisture), lowering or increasing temperature, increasing the concentration of salt or sugar or acid in foods. If you want to preserve green leafy vegetables, you have to remove the water from the leave so that micro-organisms cannot survive. You do this by drying the green leaves till all the moisture evapourates.
- *Inactivating Enzymes*: Enzymes found in foods can be inactivated by changing their conditions such as temperature and moisture, when you preserve peas, one of the methods of preservations is to put them for a few minutes in boiling water. This method inactivates enzymes and thus, in preserving the food.
- *Removal of Insects, Worms and Rats*: By storing foods in dry, air tight containers the insects, worms or rats are prevented from destroying it.

WHY DO WE NEED TO PRESERVE FOODS

Now that you have understood the definition of food preservation, did you think about why we should preserve foods? Can you think of a few reasons for preserving foods? Let us find an answer to this question by taking an example of one food item. Let us take the example of a fruit, say mango. There are many ways by which mangoes can be preserved. These are: juice, murraba, squash, aam papad, pulp, chutney, pickle, raw mango powder. Mango is a summer fruit and grows in large quantities during the months of April to August. Different varieties of mango are grown in different parts of our country. Usually all the quantity grown in a region cannot be consumed by the people staying there as there is always an excess.What does the farmer do with this excess quantity?

He makes arrangements to transport the excess quantity to regions where either mango is not grown or where that particular variety of mango is not available. If he does not do this, the excess produce will rot and go waste. The farmer will then lose money. There is still some quantity which is left after the fresh fruit is consumed by the people. It is this quantity which has to be preserved for consumption during the months when mango is not available. Preservation of foods is done during the months when food is available in large quantity and therefore at low cost:

- One of the important reasons for preserving foods is to take care of the excess produce. There are many other reasons for preserving foods. Let us learn about these.
- The second reason for preserving foods is that they add variety to our meals. Have you ever got tired of eating the same vegetables which are in season? Is it not nice to eat peas when they are either very expensive in the market or are not available ? Eating cauliflower in pulav or cauliflower vegetable during the summer months adds to the interest in meals. In the same way, eating some chatni, papad or pickle along with the meals adds to the variety. Preserving foods when they are in season makes this possible.
- Reaches areas where the food item is not grown. In some areas of Rajasthan which are desert areas and in Himalayan regions that are covered with snow most of the time, very few foods can be grown. Availability of some preserved foods can add to the variety and nutritive value of meals. For example inclusion of dehydrated peas, green leafy vegetables, canned fruits etc., in the meals is a good idea in such areas.
- Makes transportation and storage of foods easier. Preservation of foods usually reduces bulk. This makes their transportation and storage easier since it requires less space. For example, if you dry green leafy vegetables such as mint, methi, corriander, etc., their weight and volume reduces, thus making their storage easy.

WHY DOES FOOD GET SPOILT

The definition of food preservation states that preservation is keeping food in such a state that they do not get spoilt for a long period. Before we look at the reasons of food spoilage, let us understand, when is a food spoilt. When you keep bread outside the refrigerator for few days, a spongy growth is seen on it, which may be white, green on black in colour.

The bread thus gets spoilt due to growth of mould and becomes unfit for consumption. Likewise, if cooked dal or vegetables is left outside for sometime, it develops a bad smell and bubbles due to fermentation.

The dal and vegetables are thus spoilt and cannot be eaten. Can you now say when is a food spoilt? Food is said to be spoilt if there is rotting, *i.e.*, bad smell, fermentation, *i.e.*, bubbles/gas in the food or mould *i.e.*, spongy growth on the foodstuff.

Formation of soft spots or soft brown spots on fruits and vegetables is also food spoilage. Why do foods get spoilt? If you know the reasons of food spoilage, you can remove these conditions while preserving food items.

Foods get spoilt mainly due to the presence of micro-organisms, enzymes (present in foods), insects, worms, and rats.

Presence of Micro-organisms

Micro means small. Micro-organisms are very small organisms which cannot be easily seen. Micro-organisms spoil food items when the condition for their growth are appropriate.

What are these appropriate conditions? Like all living beings micro-organisms require air, moisture, right temperature and food to grow and multiply.

The situations which provide appropriate conditions for growth of micro-organisms, can be listed as:

- Food having high moisture content.
- Air around the food containing micro-organisms.
- Foods kept for a long time at room temperature.
- Skin of fruits and vegetables getting damaged, thus exposing the food to micro-organisms.
- Foods with low salt, sugar or acid content.

Presence of Enzymes

Enzymes are chemical susbtances found in all plants and animals. Are enzymes harmful to foods? No, enzymes help in ripening of fruits and vegetables.

A raw green mango after a few days becomes sweet in taste and yellow in colour due to the enzymes action.

What will happen if you keep this yellow, ripe mango for a few more days? It will become soft, develop black spots and will start smelling bad. This is due to continued action of enzymes. No one likes to eat such as over ripe, spoilt mango. You know that even when the skin of fruits is not cut or damaged, it gets spoilt. This is due to enzyme action.

PRESERVING FOODS

Unless food is preserved in some manner, it begins to spoil soon after it is harvested or slaughtered. The main methods of food preservation that will keep food safe are canning, freezing and drying.

Preserving food at home means having:

- An abundant supply of a variety of foods when the fresh products aren't readily available;
- Specialties such as strawberry-fig preserves or green tomato relish that can't always be purchased;
- The satisfaction of preserving foods yourself.

However, preserving food at home may not save you money, depending on costs of buying or raising the food, the needed equipment, energy and time.

CANNING

Canning is the process in which foods are placed in jars or cans and heated to a temperature that destroys microorganisms and inactivates enzymes. This heating and later cooling forms a vacuum seal. The vacuum seal prevents other microorganisms from re-contaminating food within the jar or can. High-acid foods such as fruits and acidified tomatoes can be processed or "canned" in boiling water, while low-acid vegetables and meats must be processed in a pressure canner at 240°F.

Pickling

Pickling is another form of canning. Pickled products have an increased acidity that makes it difficult for most bacteria to grow. Pickled products are also heated in jars at boiling temperatures to destroy other microorganisms present and form a vacuum in the jar.

Jams and Jellies

Jams and jellies have a very high sugar content. The sugar binds with the liquid present making it difficult for microorganisms to grow. To prevent surface contamination after the product is made and possible yeast or mold growth jams and jellies are canned, frozen or refrigerated.

FREEZING

Freezing reduces the temperature of the food so that microorganisms cannot grow, yet many still live. Enzyme activity is slowed down but not stopped during freezing:

- *Enzymes in Vegetables:* These must be inactivated by blanching to prevent loss of colour, flavour and nutrients. The vegetable is exposed to boiling water or steam for a specified time and then quickly cooled in ice water to prevent cooking. Blanching is essential for top quality frozen vegetables and helps to destroy microorganisms on the surface of vegetables.

- *Enzymes in Fruits:* These can cause browning and loss of vitamin C, and are controlled by the addition of ascorbic acid.

DRYING

Drying removes most of the moisture from foods. Thus microorganisms cannot grow and enzyme action is slowed down. Dried foods should be stored in airtight containers to prevent moisture from rehydrating the products and allowing microbial growth.

UNSAFE CANNING METHODS

Open-kettle Method

This outdated method of canning is now considered unsafe. In this method, foods were heated in a kettle, then poured into jars and a lid was placed on the jar. No processing was done.

With this method there was often spoilage because bacteria, yeast and molds that contaminated the foods when the jars were filled were not killed by further processing. The growth of these microorganisms, in addition to spoiling the food, often caused any lids that did seal to later come unsealed. This method resulted in a very real danger of botulism.

Steam Canning

This is a newer method of canning that is not considered safe at this time. The jars are heated by steam. However, safe processing times have not been developed and steam canners are NOT recommended for either high-or low-acid foods.

Low-acid foods canned in these canners are potentially deadly because of possible botulism contamination. Also, both low-and high-acid foods are often underprocessed and therefore could spoil.

Other Unsafe Methods

Canning food in the microwave oven, electric ovens, slow cookers or crock pots can be extremely dangerous, especially with low-acid foods, and is not recommended. So-called canning powders are useless as preservatives and do not replace the need for proper heat processing.

ADJUSTMENTS FOR ALTITUDES

As altitude increases, water boils at lower temperatures. Because the lower temperatures are less effective for killing bacteria, processing time must be increased for boiling water bath canning at higher altitudes. For pressure canning, the pressure is increased. The directions for canning foods are usually for an altitude of 0 to 1000 feet. If you are canning at an altitude over 1000 feet, check for altitude adjustments needed for canning each type of food. Ask your county extension agent to help you determine your altitude or check with your local airport.

ON GUARD AGAINST SPOILAGE

Don't taste or use food that shows any kind of spoilage! Look closely at all jars before opening them. A bulging lid or leaking jar is a sign of spoilage. When you open the jar, look for other signs, such as spurting liquid and off-odours or mold.

Spoiled canned foods should be discarded in a place where they will not be eaten by humans or pets. Spoiled meats, seafood and low-acid vegetables, should be detoxified to destroy any poisons that might be present before they are discarded.

To detoxify canned low-acid foods that have been spoiled, carefully remove the lid from the jar. Be extremely careful when following these directions not to spread or come in contact with suspect food or liquid. Place the open jar(s) of food and the lids in a saucepot. Add enough hot water to cover the jar(s). Boil for 30 minutes and then cool. Drain water and dispose of food and lids. The jars may be reused. Improperly canned low-acid foods can contain botulinum toxin without showing signs of spoilage. Low-acid foods are considered improperly canned and not safe if any of the following are true:

- The food was NOT processed in a pressure canner.
- The gauge of the canner was inaccurate.
- Up-to-date researched processing times and pressures were NOT used for the size of the jar, style of pack and kind of food being processed.
- Ingredients were added that were NOT in an approved recipe.
- Proportions of ingredients were CHANGED from the original approved recipe.
- The processing time and pressure were NOT correct for the altitude at which the food was canned.

Because improperly canned low-acid foods can contain botulinum toxin without showing signs of spoilage, they should also be detoxified and then discarded. Surfaces that come in contact with spoiled or questionable food should be cleaned with a solution of 1 part chlorine bleach to 5 parts water. Wet the surface with this solution and let stand 5 minutes before rinsing.

PRESERVING FOOD FOR SPECIAL DIETS

It is possible to preserve food at home for people who are watching their salt or sugar intakes. Simply use methods that don't require sugar or salt for their preservative properties. If preserving food for someone trying to cut sugar consumption, the food may be dried, or fruits may be canned or frozen in water or juice instead of a sugar syrup. The sugar specified in canning and freezing is only needed for its effects on flavour and texture. There are also many pickle recipes that call for little or no sugar.

Sugar is the major means by which most jams, jellies and preserves are safely preserved. These products are not suitable for a person on a low-sugar diet. However, there are special low-methoxly pectins and recipes designed to make jams and jellies with no added sugar. It is also possible to preserve food at home for people who are watching their salt intake. The salt used in canning, freezing or drying foods is used only for its flavour or colour protection quality. Pickling, especially when fermentation is involved, usually requires salt for preservative effect. Read the recipe. If it calls for a large quantity of salt, this cannot be eliminated or replaced with a salt substitute. However, if the recipe calls for a large amount of vinegar and only a teaspoon or two of salt for flavour, the salt can be left out.

TWENTIETH-CENTURY TRENDS IN FOOD PRESERVING

To set the story in a context: before the easy-to-purchase and relatively cheap food supplies of the second half of the nineteenth century, skills in the preservation of food were essential to survival. Such skills were, arguably, the most important part of the domestic economy: conserving and preserving and storing the fruits, vegetables and meats from a period of harvest or slaughter through a period of scarcity. The pattern of feast or famine was after all dependent upon unavoidable seasonal constraint on the supply of food and on the economic ability to obtain it.

Underlying all the evidence from surviving printed books, the basic necessities of sustenance depended upon a good knowledge of how to store foods and how to dry them, neither method involving any additional expenditure, and then upon salting, and to a smaller extent potting. Preserving with sugar and alcohol was, for centuries, too costly for general domestic use, and the use of vinegar, particularly in pickles and relishes, was largely for making condiments rather than preserving foods.

Knowledge of preserving techniques was so important that it must have been part of the basic training of most people up until the early nineteenth century at least, and was probably orally transmitted. In choosing to look at preserving and changing attitudes towards it from the evidence of printed books, I am taking a particular perspective that works under a number of constraints which can be touched upon by trying to respond to the question: why did people see a need to write down these accounts rather than simply transmit them orally? One reason may have been the need to respond to entirely new social events, such as the shift from a largely rural to a concentrated urban population, or to the adulteration of foodstuffs, or the effect of changes in architecture. Another reason was the need to inform people about contemporary discoveries of new techniques and preserving agents.

A third and highly problematic reason must have been the need to adapt these essential skills to the changing activities of men and women in the household, adapt them from the more integrated domestic scene of pre-1600 to the separate and relatively isolated work of the housewife, the housekeeper and the servant. Until the early nineteenth century each of these reasons, and they are only a selected few, uses the print medium to tell a specifically middle-class audience about something new, and in this sense looking at evidence from printed books yields an atypical picture of what was happening even for the middle classes.

At the same time these books do indicate a considerable amount about the construction of domestic social activity in the middle classes from the Renaissance to the coming of industrialisation. Where the picture becomes far more complicated is during the nineteenth and twentieth centuries when the print medium reaches out to a vastly enlarged audience, and we begin to see preserving skills presented in radically different ways.

The main part of this discussion will explore the reasons behind the dearth of preserving information in most domestic cookery books from the 1830s, and its reintroduction into some groups of books from the 1880s. The chapter suggests that skills which had been necessary to survival because they ensured a supply of food, whether to the agrarian or the new family domestic economy, were, for a number of reasons, simply made redundant in the early nineteenth century. What becomes of interest is why people then chose to reintroduce these skills and what reasons they had for doing so.

The books that survive from the sixteenth to early nineteenth centuries are, for the most part, substantial cared-for products that represent a relatively affluent readership. Simply due to the way books were produced during this period, much ephemeral writing aimed at labouring people has not survived, and we have few examples to indicate what might have been the concerns of such an audience.

Of the extant books concerning food from the sixteenth to mid-seventeenth centuries, most address the courtly lady or the gentlewoman emerging from the newly landed gentry; and most present preserving skills as an important domestic responsibility with one or two writers treating the field, superficially at least, as a lady-like hobby.

The earliest cookery books were concerned with sugar cookery, much of it directed to the preserving and conserving qualities of this relatively new foodstuff which was suddenly coming into Europe in large quantities. Other aspects of preserving were primarily to do with vinegars, and these sections were often combined into books with sections on medicine or cosmetics and brewing. By the 1630s, sections on preserving were increasingly often included in books about cookery, although the two sections were kept quite separate. Indeed one of the reasons preserving was considered a suitable activity for an aristocratic lady may well have been its separation from mundane cookery, as well as the prestige it carried with it of a knowledge essential to housekeeping. Partly, one suspects, in an attempt to imitate what was thought to be the life-style of aristocratic women, several late seventeenth and early eighteenth century books like *The Accomplished Female Instructor* (1704) aim to teach conduct, preserving, conserving, household receipts and medicines but not cookery.

After the Commonwealth period and particularly after 1695, books on food were aimed more and more at burgeoning middle-classes. The books that survive indicate that an understanding of preserving skills was still considered essential to the activity of the household. It was the fundamental knowledge upon which planning ahead, budgeting and responsible household management were founded. But the kind of household in question, and particularly the work of women within it, was becoming quite different from that of the late Renaissance.

An example of how the household was changing can be gained from a comparison of Gervase Markham's *Country Contentments* (1615) with Eliza Smith's *The Complete Housewife* (1727). Markham delineates the duties that fall to the woman of a household but also outlines their interconnection with the whole range of domestic responsibilities.

Furthermore, it is clear that Markham's woman of the household would not have been working alone. In contrast, Eliza Smith concentrated on preserving, cookery and brewing as specifically the housewife's or gentlewoman's province, and includes seasonal bills of fare, table-settings and medical recipes. It is not totally clear, but the impression is given that Smith's housewife is not working with a group of people although she may have had servants.

Markham's woman of the house is working within a rural economy, firmly seasonal, and she survives into the eighteenth century, mainly in the form of the housekeeper to large landed houses and partly in the new profession of innkeeper. Smith's housewife, still aware of the seasons, is an urban creature dependent upon street markets, whose successors become less and less gentlewomen and more plain housewives to whom preserving skills are, as in Penelope Bradshaw's *Family Companion* (1750) or Susannah Carter's *The Frugal Housewife* (1765) or E. Spencer's *The Modern Cook; and Frugal Housewife's Complete Guide* (1782), a matter of domestic management and frugal housekeeping. Until the early nineteenth century nearly all women would have had to know something about preserving food. Apart from alerting women to the professions of housekeeper and innkeeper and popularising individual writers, the main reason for printed books on the topic must lie partly in the changing social structure of the middle-class domestic household, which gradually placed all the responsibility for these skills onto the single housewife — although she may have directed servants to help her — and partly in the market-based economy which the housewife had to learn to use frugally.

Later in the eighteenth century a better understanding about the principles of preserving food, with a new technology applying the understanding and with commercial outlets exploiting the product, fundamentally undermined the need for such skills in a domestic setting aimed at frugality and in doing so fundamentally altered the activity of the housewife: she no longer preserves but buys, she no longer produces for herself but consumes products made for her.

More than this, activities that would have occupied a large part of domestic budgeting and household management no longer needed to be carried out. In order to grasp the far-reaching effects of changes in the technology of preserving I would like to summarise some background dealt with in greater detail elsewhere in this book.

Preserving techniques in Britain until the early nineteenth century, in descending importance to the majority and in ascending cost, included drying, salting, potting, vinegar pickling, sugar-bottling or candying, and conserving in alcohol. As with all preserving techniques, they depend either on the addition of a preservative or upon the deprivation of the rotting agent which is usually either moisture, air or heat. Sometimes the former affects the latter.

The main aim of each technique was to keep the foodstuff edible until such time as it was to be eaten, although the process would have produced flavours and textures which came to be appreciated for themselves. The most widely written-about techniques in these early books for the middle-classes discuss the use of salt and sugar. The story of nineteenth century preserving technology is mainly about how the use of sugar in bottling came to be superseded by canning, and the use of salt in dry-salting, brining and pickling came to be superseded also by canning but more importantly by refrigeration. In the late medieval period sugar was a very expensive product. It was soon recognised that it could be made to go further by diluting it with water into a syrup.

Although this lessened its preserving qualities, the bottles that held the syrup served another function by excluding air. The exclusion of air was an important principle that interested many early scientists such as Robert Boyle, and people who wrote books with sections on preserving often implied and sometimes explained that bottling was the main technique allowing for economising on sugar.

Yet even in the eighteenth century there was an awareness of further complexities, and writers advise on the preference for narrow-necked jars and warn their readers not to put their fingers, or utensils that have been used for other foods on the table, into jars of preserves. Until the late eighteenth century, most recipes concentrate on excluding air from coming in: there are elabourate combinations of lids, paper, brandy-paper, leather, bladder, cork, cork and resin — and excluding materials — fat (preferably mutton because of its low setting point), oil, or juice (for example lemon).

But it was gradually recognised, particularly following Priestley's work with oxygen in the 1760s, that another problem was the 'air' already in the jar, in the very mixture to be preserved. This problem was referred to as 'oxidation' and was the main concern of the next advance in technology.

Due to a sugar shortage during the Napoleonic Wars, the post-revolutionary government established a prize to encourage new work on the preservation of food in bottles. In 1808 Saddington proposed, to the English Society for Arts, the boiling and scalding of bottles as a solution; and in 1810 Nicolas Appert finally won the large money prize from the French government for his careful organisation of heated bottles with a champagne-type seal.

The results of Appert's work were studied by the chemist Guy-Lussac, who concluded that it was the oxygen being driven out of the bottle during heating, and then prevented from re-entering, which resulted in success and the prevention of oxidation. At precisely this time, again with the practicalities of army provender in mind, Peter Durand in England took out a patent on sealing food in cans. Canned food was taken on Parry's arctic explorations in 1814-16, and came into its own during the Crimean War.

By the 1830s the principles were better understood and the technology more precisely applied so that there was a reliable method for producing bottled goods, and the same could be said for canned goods by the 1850s. But the products had to reach the buying household. This they did through the explosion in retail trade outlets or shops occurring in most urban centres from the 1840s, and eventually all but supplanting the street markets. The products needed the shops and the shops needed the products. Markets were increasingly less central to the dominant urban economy and the housewife was urged instead into shop-bought bottled and canned goods by a growing advertising industry based on the packaging of the goods.

Evidence that the housewife responded does seem to be provided by the cookery books of the early nineteenth century. Elizabeth Raffald's *The Experienced English House-keeper* which was published several times between 1769 and 1796, has a large and consistent section on preserving. In contrast, an equally popular nineteenth century book, *A New System of Domestic Cookery* by Maria Eliza Rundell, includes a large number of preserving recipes in the early edition of 1807, but by 1819 there are far fewer.

Some recipes were shifted to a section specifically on sweetmeats. Also of possible significance, whereas the writer/editor of the 1807 edition explicitly stated at the start of the sweetmeat section that these recipes were less important for private families since such things could be bought at far less expense than they could be made, the identical statement is transposed to the preserving section in the edition of 1819. By the middle of the nineteenth century, so few cookery books contain a substantial section on preserving that the exceptions to this rule are particularly informative. In Eliza Acton's *Modern Cookery* of 1845 there are detailed instructions for preserving fruit, but we have to remember that this is the same Acton who was a working journalist, concerned with issue of public welfare, and who was to write *The English Bread Book* (1857), specifically to address the problem of the adulteration of shop-bought bread by encouraging people to make their own.

By the 1868 edition of *Modern Cookery,* Acton feels the need to make the same intentions clear for her section on preserving and scathingly criticises the 'unwholesome [preserved] fruit vended and consumed in very large quantities' by the shop-buying public. Acton's stress on the 'wholesome' is a significant precursor of the direction that preserving recipes will take when they re-enter cookery books at the end of the nineteenth century. No longer can the housewife claim to be frugal when she uses preserving skills, but she can claim to produce more nutritious and healthy food. The same argument, however, did not apply to salting.

Techniques of preserving that used salt were far more widespread than those using the more expensive sugar, and much older in cultural practice. Although eggs and some vegetables, particularly green beans, were salted, the main foodstuff preserved in this way was meat. Beef cattle in particular were slaughtered wholesale in the autumn because there was rarely enough hay to keep them through the winter, so the meat had to be preserved.

Brining and dry-salting must have been such a central part of general life that it is not surprising that there are few recipes in the early cookery books, despite the fact that with the imposition of taxes on salt from 1785 to 1825, frugality with salt must have been important.

By the end of the eighteenth century, most books included some recipes, for example there is almost always one for Westphalia ham and salt beef, and from the early nineteenth century there is an increasing number of recipes for foreign salted foods like polonis and salamis, presumably in response to wider continental travel of private families. But these recipes, too, disappear from most domestic cookery books after the mid-century, only surfacing in a few trade receipt books for professional shop-keepers like James Robinson's *The Whole Art of Curing, Pickling and Preserving*. This work is addressed to the trades, particularly the fishmongers, but includes (unacknowledged) many of Rundell's recipes: the line between commercial and domestic preserving was not yet clearly drawn.

It appears that the main reason that salting recipes fade from cookery books, apart of course from the advent of the can, is the introduction both of ice-cooled and of refrigerated boxes during the 1830s. The use of cold to preserve food had fascinated the English from Bacon's experiments in 1626, to Pepys' amazed discussion in the 1670s on the frozen Baltic chickens, to the later 1799 discovery of the frozen mammoths in Russia. Since the Renaissance, several great country houses in Britain had had ice-houses, but until Rundell's advice in 1807 on unfreezing food slowly, there were few remarks in general domestic books.

From the early 1800s the trade in ice grew from the first imports from Norway in 1822, to the artificial production of ice in the 1860s. At the same time Jacob Perkins, an American engineer working in Britain, developed the ice-cooled refrigerator and produced it for sale between 1835 and 1870. From the 1860s meat was available in cans, and from the 1880s cheap imports from Australia and New Zealand were being imported in cans and by refrigerated ships.

A secondary reason for the loss of salting recipes was a growing concern with the nutritional value of preserved foods. Michael Donovan, that centre of sense, comments that 'The object is... to preserve, as much as possible, the nutritiousness of food and its salubrity, and to prevent its doing actual injury to health...'. Just as the French army was concerned with dependable food supply during the Napoleonic Wars, so the British army launched intensive study into methods of food preserving during the 1840s and 1850s leading up to the Crimean War.

Food and its Adulterations compiled by A. Hill Halsall in 1855 lists a series of tests conducted by and for the army in ways of keeping meat without salting it: not only canning and drying but also smoking, packing in tin foil, vacuum packing and producing extracts of vegetables and meat. In 1859 Beeton states that the action of salt on meat decreases its nutritive value, and by 1870 William Tegetmeier is advising against the consumption of much if any salted meat.

There is some evidence that American trends in health foods also began to have some influence on British attitudes towards eating at the end of the nineteenth century. Russell Trall's *The New Hydropathic Cookbook,* published in England from 1883, cites salt meat as the prime example of bad food. Also during this period there was the vigorous commercial marketing of Liebig's 'Extract of Meat' for domestic use, by the LEMCO organisation whose advertising suggested that salting meat was deleterious to health. Indeed, apart from one or two recipes apparently included to satisfy those who craved the taste of salted meat, and a brief resurgence from 1918 to the year 1925 (when the Public Health Preservation in Food Act actively promoted domestic refrigeration), only a few recipes for salting meat occur in British cookery books until Jane Grigson's *Charcuterie* of 1967.

Some of this account may help us partly to understand the dearth of preserving recipes in domestic cookery books from the 1830s to the 1880s. What is more difficult to assess is why any such recipes should have reappeared at all. From the 1850s onward, the audience for printed cookery books expanded considerably, and exploded during the 1870s and 80s on the back of a proliferating periodical publishing industry, which was creating markets and responding to the demands for a wide range of general topics accessible to the reading publics that were defined in the aftermath of the Education Acts of 1867 and 1870. At the same time cookery books continued their progress into specialised genres, and focused not only on narrower topics but also on class divisions. Alexis Soyer's *The Gastronomic Regenerator* (upper-class, recherche dishes), *The Modern Housewife* (middle-class) and *Shilling* Cookery (artisan) is but one example of a specifically class-orientated approach common to many writers of the mid-century.

But for high or low, rich or poor, few of these books contained advice on preserving although several contained advice on how to use canned goods, for instance Emily de Vere Matthew on *Tinned Meats* (1887), and Jane Panton's *From Kitchen to Garret* (1888) which advised on the use of cheap canned and frozen New Zealand meat. One exception which *may* indicate a group of "lost" books is the publication from the Labourer's Friend Society *A Second Series of Useful Hints for Labourers* aimed at the artisan with an allottment and containing advice on how to salt your pig.

However, in the explosion of the 1870s and 1880s a distinct pattern emerges: books for working-class women do not contain recipes for preserving while those for middle-class women, and slightly later also for wives of artisans, do. *The Official Handbook for the Training School for Cookery,* published from 1888 for many years, contains recipes for using canned meat, one recipe for pickling meat and one for pickling cabbage.

The foremost reason for the lack of preserving recipes is economic. The wider ranging household management and planning requirements of preserving of any kind imply having the time to do it, the money to buy ingredients in bulk as well as preserving agents and equipment, as well as, usually, a source of heat and always space for storage.

Most of the urban working-class population would not have had easy access to all, or indeed any of these requirements. The very concept of there being periods of plenty during which people save up against periods of famine is not part of an industrial working pattern. Neither the school text-books written for this audience nor the more general books of advice to these housewives, consider preserving as an important skill. Indeed W. Tegetmeier's *A Manual of Domestic Economy* account of working-class diet assumes that any fruit and vegetables are a luxury to this class, let alone being available in quantities enough to preserve.

And fifty years later C. H. Senn's *Popular Cookery* of 1920 assumes that its readers buy in any preserved foods they need from shops — although his *Practical Gastronomy* for the middle-class housewife contains a range of recipes for conserves and pickles. Contrary to some current speculations, there is growing evidence not only from the cookery books but also from recent research in social history, that cans and can-openers were accessible to and used by the working-class population of the late nineteenth century.

Unlike schools for working-class girls, middle-class schools did not teach cookery, so the text-books on general food preparation and management for these readers were usually written for specific Cookery Schools. Apart from these books this audience was also addressed by practical household management books which assumed that the housewife had at least one servant, by textbooks written for women attending the National Training Schools of Cookery in order to teach domestic economy in schools, and by conduct or etiquette books whose purpose was to provide information about what needed to be done in your household but did not expect you to do it yourself.

There are several strands offered by these books when they include recipes for preserving, but the most important is that of nutrition. To understand the influence of this strand, it is useful to look again at changes in the understanding of the principles of food preservation. Until the early 1800s, the scientific and technological attention to preserving had focused on preventing the foods from rotting for immediate reasons of economy and frugality.

Nineteenth-century scientific advances document a shift in focus away from frugality to nutrition: a move that closely parallels and must be interconnected with the same shift in the cookery books of the period, for preceding the resurgence of preserving recipes in books for a middle class audience are the few books on vegetarianism or 'health' foods, usually beginning in the United States, directed with prophetic fervour to a more wholesome diet. These isolated books from the 1840s to the 1880s are being produced at the same time that Pasteur and Tyndall are doing their separate but related work on the effects of yeasts, moulds and bacteria on the deterioration of food.

Evidence that air carried not only 'oxidising' agents but also unseen organisms that rotted food, led to even more efficient systems for bottling and canning. However Britain, possibly because of its concentrated urban, industralised and shop-centred domestic economy, was slow to respond. Despite the success of George Fowler's patented bottling system during the 1890s, even in 1916 the Royal Horticultural Society is complaining about the need to import good bottles from the Continent, and canning was never a serious proposition for domestic preservation even in National Training School text-books, although some books from the United States containing substantial instructions on canning were published here from 1890 to 1920 and there was certainly a domestic canning machine on the market.

At the turn of the century scientific attention began to focus on the newly-discovered presence of enzymes, along with an extended understanding of fermentation processes and an appreciation of the role of vitamins in body chemistry. For food preservation, the end result of this attention was a proliferation of suggestions for preserving food with as little heat as possible.

Concern for the role that enzymes played in determining the taste of food as well as concern for the delicate biochemical balance maintained at body temperatures, meant the recognition that heating foods to preserve them in bottles was destroying their nutritional and gastronomic value.

One solution was a short-lived attempt to promote the pressure cooking of cans for brief time periods, and another was the extensive and helpful experiment by the Royal Horticultural Society on the shortest effective heating times and lowest effective temperature for bottling which led to a classic book on bottling which has been adapted and revised for nearly a century: *Fruit Bottling*. But, ironically, the primary response to the deleterious effects of heat was to consider what could be 'added' to the environment of the food to stop rotting agents getting to it while preserving its natural processes. In the balance were two processes.

The first concentrated on providing a surrounding unfriendly to bacteria and moulds, and here chemical additives came into their own. In 1859 Isabella Beeton, among a few others, had advocated using alum in bottling fruit in order to increase the dependability of preserving, but from the late 1800s more and more chemicals were added for this purpose, the most common being sodium sulphite.

The second process to which people at that time paid increasing attention, was that these additives might not only prevent the entry of external rotting agents but might also slow down, possibly by altering it, the natural maturation of the foods themselves.

It is difficult to differentiate between the two processes and the second clearly changes the nutritional value and probably also the taste of the foodstuff. Worry about exactly what these additives were doing became so widespread, as use of them became more and more ubiquitous, that the government had to control them, and in 1925 published the Public Health Preservation in Food Act, which is in effect an early listing of allowable and prohibited E-numbers.

The focus on the wholesome and the nutritious by scientists and technologists, was filtered out into books for a middle-class audience primarily through the text-books for training domestic economy teachers. These text-books, initially appearing in the 1870s and 1880s, claim for those who study them a serious area of learning. In most the pedagogic tone is based on the rhetoric of modern science: fact, proof, method and explanation.

It has been suggested elsewhere, and I think it is underwritten by these books, that domestic economy teachers were trying to define a field of academically respectable study in order to justify the serious status of their work; and since there were more obviously scientific aspects in preserving than elsewhere in cookery, it was a clear field for emphasis in their training.

The kind of book produced for these training schools, such as Catherine Buckton's *Food and Home Cookery* or E. G. Mann's *Domestic Science Manuals* among very many others, drew on an earlier tradition of male writers who had worked in the technical schools of the 1850s to 1870s. John Buckmaster's lectures on cookery were brought together into *Buckmaster's* Cookery and William Mattieu Williams produced The Chemistry of Cookery. Both writers were widely influential, particularly Buckmaster who lectured at the National Training School for Cookery in London, and both focused on science as the basis for domestic cookery, Williams stating that

The kitchen is a chemical labouratory in which are conducted a number of chemical processes by which our food is converted from its crude state to a condition more suitable for digestion and nutrition, and made more agreeable to the palate.

In turn, Buckton and her colleagues in their text-books for training teachers, are concerned to discuss oxidation, fermentation, fibres, yeasts, moulds, bacteria, pasteurisation, sterilisation and so on; and they do so largely in the context of the preservation of wholesome and nutritious food.

When the students of the training schools went on to write domestic cookery books of their own at the turn of the century, as many of them did, preserving once more found a place, again firmly in terms of nutrition and wholesomeness. For example there is M. Fairclough's *The Ideal Cookery Book,* written for her cookery school, and containing a large section on preserving.

The domestic reader had books such as *The Housewife's Cookery Book* published in 1920 but in effect an update of *Warne's Model Cookery* from 1871, which begins with the claim that 'The science of cookery is a knowledge of the choice of food and food materials, for just as an engine requires food to enable it to work, so the body requires certain foodstuffs to keep it in working order', yet continues much as the 1871 edition, but with the addition of a substantial section on bottling.

Less pretentious were the many specialised books focusing on preserves such as Helen Souter's *Aunt Kate's Jams and Jellies Book* or Rose Brown's *Pastry and Preserves* or S. Beaty Pownall's *Queen Cookery: Pickles and Preserves*. These writers also contributed to a small group of books aimed at teaching the newly impoverished genteel lady how to make delicacies that one could no longer afford to buy: an interesting index to the fact that economy could be reintroduced as an element.

Furthermore, they also produced a well-defined but still limited group of books aimed at the upper-middle class lady with time on her hands and searching for something to do. Just as this lady was usually expected to be able to produce at least one meal on her own, a star turn, for the delight of her husband, she could also turn to the making of preserves as a delicate hobby. Like the newly-emerging monied woman of the early seventeenth century to whom she was occasionally compared, the late nineteenth-century lady could not be involved in mundane cookery but could take on the role of producing elegant gifts, acting the Lady Bountiful. These Lady Bountiful books formed a curious partnership with the other main promoter of new preserving skills: the Royal Horticultural Society. Possibly due to the Allotment Act of 1887 which made it legally necessary for all local authorities to provide allotments, large numbers of people were growing their own vegetables and fruits, and from 1910 the Royal Horticultural Society was awarding prizes and medals for preserved fruits and vegetables.

In 1916 the society published Wilks' *Fruit Bottling,* its flagship guide to preserving. However, the response from individual members of the Horticultural Society began earlier with for example the series of articles by May Crooke in the 1905 *Farm and Garden* on the small holdings commission, or the Stoney's *A Simple Method for Bottling Fruit at Home* or Edith Bradley's *The Book of Fruit Bottling*. This last contains an introduction by Wilks in which he comments that fruit bottling had gone downhill from the 1840s when he was a boy to its nadir in 1885, when 'I doubt whether there were a dozen ladies in the land who did their own bottling and preserving'.

These books arrived just in time to lay a basis for response to the demands of World War I. Ernest Oldmeadow straightforwardly entitles his contribution: *Home Cookery in War-Time*, and resuscitates many eighteenth-century preserving methods such as the use of mutton fat to seal jars, in an attempt at economy. While the Great War did remind people that domestic preserving could save money, the stress was still on health. Louise Andrea's *Home Bottling Drying and Preserving* emphasises the conservation of 'nutritive value' and the 'wholesome', but in the light of war-time experience she proclaims 'Empty jars can be put to splendid use; if allowed to remain empty they are voiceless but eloquent reproaches'; indeed 'Empty jars are slackers'. By the end of the war Bristol University had established a Fruit and Vegetable Preservation Research Station.

After the war there was another wave of books specifically on preserves and of general books containing substantial sections on preserves. Among the former are the Banks's *Fruit and Vegetable Bottling,* the Royal Horticultural Society's update on Wilks' *Fruit-Bottling,* and the closely related *Domestic Preservation of Fruit and Vegetables*, edited by M. L. Adams for the Ministry of Agriculture. Among the latter are found the Good Housekeeping Institute's newly influential guides.

The Good Housekeeping Institute's director, D. Cottington Taylor, also published *Frigidaire Recipes* which is by and large cooking with refrigerated foods in 1930, and by 1938 the Canned Foods Advisory Bureau employed Janet Bond to write *Janet Bond's Book:* A *Practical Guide to the Use of Canned Foods,* in which she claims that canned goods not only preserve more vitamins than domestic preserving, but have also arrested the decline in fruit culture, freed women from the kitchen into careers and acted as the guardian of the Nation's

Health. But 1939 and World War II shifted the emphasis firmly back on to preserving within the home and the 1940s saw a whole series of guides to effective methods, led by Crang and Mason's revision of Adams' book, for the Ministry of Agriculture, under the same title *Domestic Preservation of Fruit and Vegetables*. The book begins resolutely on the first page saying that: the object of preservation is to take food at its point of maximum palatability and nutritive value and keep it at this stage... a study of these changes has shown that they are due partly to the action of enzymes in the food, partly to the growth of micro organisms in contaminating them.

It proceeds with instructions for jams, jellies, marmalade, cheeses and butters, bottling, canning, deep-freezing, chemicals, syrups, candying, preserving vegetables, vinegars, pickles, chutneys, sauces, drying, salting beans and storing. Crang and Mason's work, apart from the sections on deep freezing, could have been written before World War I, but it includes a rather different audience, the working class, in its address.

Most of these books on preserving from the 1880s to the 1940s were resolutely directed towards a middle- and upper middle class audience, and the class specificity raises a difficult problem. Ever since domestic preserving as an essential for survival was superseded by shops and their cheap goods, it had occupied a tenuous place in cookery books. Exactly because it moved out of the mundane, or those things essential for everyday survival, it had nothing to fall back on to prove its necessity or relevance. Preserving, while attempts were being made to reclaim it in terms of nutrition, became fundamentally superfluous to daily needs.

In a curious way the situation focuses on the fact that during the nineteenth century preserving in any but the most basic sense was beyond the poor, and then the working-classes: they were too impoverished to practice frugality and thereafter too impoverished to practice nutrition. It is ironic that while the Napoleonic and Crimean Wars precipitated the processes of domestic preserving out of women's hands, World War I and particularly World War II, partially put it back.

The events related to these two twentieth-century wars also broadened the constituency for preserving by changing food supply routes, by making preserving once again essential to survival; and, possibly most importantly, because there had been a radical break in the oral history of preserving techniques, from say 1830 or 1840 to 1910, the surge of printed book instructions at the turn of the century were not just addressed to a privileged audience but to a much wider range of reader.

An introduction to the basics went hand in hand with an introduction to condiments, relishes and conserves that were not strictly essential. General cookery books from the immediate post-1939 period invariably contain a section on preserving techniques, although these largely diminish to conserves and pickles by the end of the 50s. Why do people continue to preserve on any scale? This is an extremely difficult question to answer. After all we do not technically need to do so for survival or nutrition, and there are many cheerful commercially-produced gift bottles around.

For certain generations, in which I would include my own, there is a nostalgic remembrance of childhood post-war rationing days when it was necessary to bottle and jam, to harvest in plenty, to put up against want and to waste not: then as now it was often a communal and social activity.

For some, there is the confidence of knowing exactly what has gone into the foodstuff: home preserving is the only sure way of evading major additives and of controlling sugar content, and so on.

For many, there is the delight in the activity of cooking and in the beauty of the product, neither of which is strictly necessary. Possibly for most people preserving has become a kind of signature.

Making our own preserves, conserves and pickles allows us to vary the flavour and taste of a recipe that is frequently traditional, and to recognise quite precisely how the network of food distribution and supply, quality and quantity, changes from year to year.

The products are also often communally distributed, so that every year you wait with anticipation for one friend's pickle and with dread for another's jam; they are the staple of countless school sales and fund-raising fairs. Preserving brings the general domestic cook closer to the subtleties of cooking than our mundane cooking normally allows, and in our culture it is looked upon primarily as a leisure activity. How long it will remain so is unclear.

CHEMICAL FOOD PRESERVATION

There are three classes of chemical preservatives commonly used in foods:

- Benzoates (such as sodium benzoate),
- Nitrites (such as sodium nitrite),
- Sulphites (such as sulphur dioxide).

If you look at the ingredient labels of different foods, you will frequently see these different types of chemicals used. Another common preservative that you will commonly see on food labels is sorbic acid. All of these chemicals either inhibit the activity of bacteria or kill the bacteria.

CHEMICAL TREATMENT

Chemical preservatives are added to kill or inhibit microorganisms in food. They may be incorporated into the foods, or only on their surface. The wrappers used to cover them may be treated, or there may be the use of gas or vapours around the food. Some chemicals may be effective on selected group of microorganisms while others on a wide variety of them. Chemical preservatives may be harmless if they are added during the storage period and are removed before the food is consumed. But, it they are consumed as such, they may be poisonous to man or animal as well as to microorganisms.

ORGANIC ACIDS AND THEIR SALTS

Several organic acids and their salts are common preservatives as they have marked microbiostatic and microbicidal actives. Benzoic acid and benzoate are used for the preservation of vegetables. Sodium benzoate is used in the preservation of jellies, jams, fruit juice and other acid foods.

Salicylic acid and salicylates are used as preservatives of fruits and vegetable in place of benzoate because the latter is considered to be deleterious to health of consumer. Scorbic acid is recommended for foods susceptible to spoilage fungi, *e.g.*, it inhibits mould growth in bread.

Wrapping material for cheese may be treated with it. It is also used in sweet pickles and for control of lactic fermentations of olives and cucumbers. Foods prepared by fermentation processes, *e.g.*, milk products, etc., are preserved mainly by lactic, acetic and propionic acids. Flavouring extracts of vanilla, lemons are preserved in 50-70 per cent alcohol as it coagulates cell proteins.

INORGANIC ACIDS AND THEIR SALTS

Most common among the inorganic acids and their salts are sodium chloride, hypochlorites, sulphurous acids and sulphites, sulphur dioxide, sodium nitrate and sodium nitrite.

Sodium Chloride

Sodium chloride produces high osmotic pressure and, therefore, causes destruction of many microorganisms by plasmolysis. It causes dehydration of food as well as microorganisms, releases disinfecting chlorine ion by ionization, reduces solubility of oxygen in the moisture, sensitises microbial cells against carbon dioxide and interferes with the action of proteolytic enzymes. These are the reasons why this common salt is used widely for preservation either directly or in brine or curing solutions.

Hypochlorites

The hypochlorous acid an effective germicide provided the organic matter content of the medium is not high. It is oxidative in its action. The commonly used forms are sodium and calcium hypochlorites. Drinking water or water used for washing foods or icing them may be dissolved with hypochlorites.

Sulphurous Acids and Sulphites

Sulphurous acid and sulphites are added to wines as preservatives. Sulphurous acid is used especially in the preservation of dry fruits. It helps retention of original colour of the preserve and inhibition of moulds more than either yeasts or bacteria. Potassium metabisulphite is used in canning.

Sulphur Dioxide

Sulphur dioxide has a bleaching effect desired in some fruits. And also suppresses the growth of yeast moulds. It is used as a gas to treat drying fruits and is also used in molasses.

Nitrates and Nitrites

Nitrates and nitrites produces an inhibitory effect on bacterial growth and used usually together in meat and fish preservation and for retension of red-colour of the meat. Nitrate is changed to nitrous acid which reacts with myoglobin to give nitric oxide myoglobin. It is the latter which gives a bright red colour to the meat making it more attractive in appearance. However, both nitrite and nitrate are poisonous if present in potable water or food products in more than minimal amounts. This is why the generous are of these chemicals as preservative in meat and fish products has been questioned.

ANTIBIOTICS

Aureomycin (chlorotetracycline) is the most commonly used antibiotic for the preservation of animal products under chilling conditions. It is extensively used for the preservation of poultry, meat, and fish. The antibiotic is applied to the surface of the fresh meat by dipping it in a solution of the antibiotic or it may be fed to the animal by mixing it with feed or water for one to several days before slaughter. Fish are treated by adding the antibiotic in the ice or water in which they are to be transported.

The indiscriminate use of antibiotics as preservatives, however, should be prevented or the antibiotics used should be such that is demobilished on cooking so that the internal flora of man using such food is constantly exposed to the effect of the antibiotic. It is important because otherwise use of antibiotics would lead to the development of the antibiotic resistant strains of microorganisms in the body. Aside from this, some individuals' sensitive to antibiotics becomes exposed constantly to allergy.

Preservation of foodstuffs has always been a necessity for a number of reasons: the durability of food is limited, numerous foodstuffs are only available during a short harvesting season, the transport routes of food or raw materials from the production site to the consumers are continuously increasing in length and the consumers in modern society characterised by division of labour and changed shopping habits increasingly insist on buying durable products. Beyond this, there are medical-hygienic efforts aimed at inhibiting the growth of micro-organisms in food. The hazard to health which many bacteria carry, has been long known. Recently, a number of fungi have been shown to form toxins during their growth on foodstuffs. There are two methods of food preservation: the physical and the chemical. The greater proportion of foodstuffs is rendered durable by physical procedures: drying, cooling, deep-freezing and heating. But chemical preservation also plays a prominent role. The use of preservatives is often combined with physical methods. The application of preservatives has a long history, such as the use of common salt, smoke or sulfur dioxide. Some of these agents, such as benzoic acid, are achievements of the last century. Others, such as propionic acid and sorbic acid, result from research during the last few decades.

The preservatives now in use have been thoroughly tested for their toxicological properties. Their use in the food industry is subject to stringent legal regulations. The consumer can be certain of not running any risk by partaking foods which contain preservatives. Methods of food preservation have been known for thousands of years. Preserved samples of food have even been discovered in ancient Greek urns.

Chemical preservatives work either as direct microbial poisons or by reducing the pH to a level of acidity that prevents the growth of microorganisms. Acetic acid, better known as vinegar, has also been used as a food preservative since ancient times. Salted, pickled or dried foods were about the only nourishment sailors were offered on long sea voyages before the invention of modern refrigeration and preservation techniques.

CHEMICALS USED TODAY

Two commonly used preservative chemicals are:

- Nitrates and nitrites that are used to preserve meats such as ham and bacon.
- Sulphites that are commonly used to prevent the browning of fruits and vegetables after they've been peeled, and to prevent fungal spoilage.

As important and useful as they are, preservatives have developed a bad name in Western societies such as Australia. Salt is now widely shunned because of its effects on blood pressure. Nitrites and sulphites can both cause asthma, nausea, vomiting and headaches in some people. For these reasons, consumers have started to demand foods containing lower levels of chemical preservatives. The potential drawback of this is the reduced length of time before conditions favour the rapid multiplication of food-poisoning agents like *Salmonella* and *Listeria*. A number of food-processing techniques have been developed to prolong the shelf-life of foods and permit a reduction in preservative levels.

NATURAL PRESERVATIVES

Scientists are putting increased efforts into the discovery and purification of natural compounds for use as safe alternatives to chemical preservatives. The new breed of protective compounds are small proteins. They are called bacteriocins and are starting to be used in a wide variety of foods. Anyone who has eaten yoghurt has been protected by bacteriocins without knowing it. Bacteriocins are produced by some good bacteria to kill competing organisms such as *Listeria monocytogenes*.

The whole bacteria that produces the bacteriocin, or the purified bacteriocin itself, can be added to foods such as soft cheeses to reduce the risk of pathogen growth. An example of protection using bacteriocins is the use of nisin in crumpets to restrict the growth of *Bacillus cereus*.

LONG-LIFE TREATMENTS

Modern technology has produced several new processing techniques for prolonging the shelf-life of perishable foods.

UHT

Ultra high temperature treatment (UHT), for example, involves the rapid heating of food to about 140°C. This temperature is maintained for a few seconds to kill bacteria. The product is then cooled rapidly and placed in sterile, airtight containers to prevent recontamination. This treatment is used commonly to produce 'long-life' milk and fruit juices. A disadvantage of the high temperature treatment is that heat-sensitive vitamins such as vitamin C are destroyed. In fruit juices the vitamins are added back after treatment.

MODIFIED ATMOSPHERE PACKAGING

As the name implies, modified atmosphere packaging (MAP) alters the mix of gases-notably oxygen, nitrogen and carbon dioxide-in the atmosphere in which foods are stored. The altered storage atmosphere can limit the growth of spoilage bacteria and fungi and extend the shelf life of food. The technique is used for bread, cakes, fruit, vegetables, meat and seafood, which are stored in special plastic bags that help maintain the atmosphere for a certain length of time.

The technique is also applied to bulk foods in especially equipped storage containers. This is how fruits such as apples can be supplied to consumers throughout the year, long after the apple harvest takes place. Although modified atmosphere packaging (MAP) limits the growth of spoilage micro-organisms, it does not slow the growth of some harmful bacteria. For this reason, MAP is usually used in conjunction with other preservation techniques such as refrigeration.

HOME CANNING FOOD PRESERVATION

THE BENEFITS OF HOME CANNING

Canning food in your own home is a safe and rewarding process that is becoming popular again as food prices soar and people realise they need to pay attention to securing their food supplies. Preserving food with home canning is an excellent way to increase your consumption of local food. Eating locally requires eating foods when they are in season, and canning allows you to capture the bounty of any particular crop in season and extend its availability throughout the year.

You can approach home canning as a hobbyist or a full-time enthusiast who stocks a sizeable percentage of his or her food supply with home preserved food. Whether you want to enjoy a couple fun weekend projects putting up jam or seriously supplement your diet, you will enjoy many personal benefits while being a good steward to the environment and supporting your local economy. And the way things are going with the global food market, you will likely save money as well, especially as time goes on.

How Preserving Food at Home Helps Us

- *Excellent Quality and Taste*: When you use quality produce and perform the canning process correctly, you will create superior products to those for sale at the supermarket. Many recipes for home canned food are delicious and literally the quality is something that money can't buy. You have to make these luscious foods yourself.
- *Control over the Ingredients*: With home canning, you will know exactly where your food is coming from. Ideal sources of produce are your own garden and fruit trees, local organic farms, and any local farm. From any of these sources you will be able to hand select your produce at the peak of ripeness. With home canning you will also reduce your exposure to Bisphenol A that lines the cans of many mass produced food products. Bisphenol A is an endocrine disruptor and people are becoming increasingly aware of its potential harm to humans.
- *Support of the Local Economy*: By directly buying produce from local growers, you are putting money into the hands of local people. Local growers love selling from their own farms or market stands because they are not at the mercy of the big commodity buyers who set prices. This also allows local growers, especially small ones, to remain profitable, which is good for the local economy.
- *Lower Your Carbon Dioxide Footprint*: Great amounts of energy are used to produce and transport the food eaten by society. Highly industrialized agriculture also relies on pesticides, herbicides, and petrochemical fertilizers. All of these things are bad for the environment and degrade the ability of soils to produce food in the future, which means greater scarcity, lower quality, and higher commodity prices. When you buy local food and can it at home, you are eliminating a huge percentage of the transport costs from burned

fuel associated with shipping food across continents. Yes, home canning requires an energy input, but it does not compare to food being trucked halfway across the country to stock a shelf in a store. Reducing the amount of food you eat from distant places reduces the amount of fuel you are causing to be burned. Also when buying local food, try to focus on those growers who use sustainable growing practices that do not poison the environment.

- *Sense of Accomplishment*: Once you begin canning food, you will be thrilled with yourself. You will feel like you did something very meaningful to your existence because you did! For most of human history most people focused a great deal of time and energy on securing their food supplies.
  We are not suggesting we all go back to digging for roots in the field, but people in general have a deep need to participate in the gathering and preparation of food. Sitting in an SUV for drive-through fast food does not satisfy. It only promotes outrageous energy consumption for low quality products.

Now that you are ready to do some work in the kitchen and participate in food preservation, you need to understand the basics of home canning. The process is a little intimidating at first, but after a couple canning projects, you will feel much more comfortable doing it.

There are safety considerations with home canning, but these are all satisfied by following the directions associated with any particular home canning project.

The manufacturers of Mason jars (which are the essential product needed for safe home canning) offer great information on how to can food at home, and many university agriculture departments and the USDA offer very reliable information on this subject as well.

Food Preservation with Home Canning Halts Spoilage

Food preservation in canning jars is accomplished by killing spoilage causing agents with heat, removing air from the food products, and sealing the jars so that air and yeasts, moulds, and bacteria cannot be reintroduced to the food.

Four Causes of Food Spoilage

- Enzymes – Destroyed at 140°F.
- Moulds – Destroyed at 140°F to 190°F.
- Yeasts – Destroyed at 140°F to 190°F.
- Bacteria – Many types of bacteria exist. The toxins produced by some bacteria are also a hazard. Bacteria and associated toxins are destroyed in heat ranges from 190°F to 240°F.

Some bacteria are very tough and resist death even at high temperatures. The toughest bacterium is *Clostridium botulinum*, whose spores cause the deadly botulism. This bacterium is killed at 190°F and its toxic spores are destroyed at 240°F. This bacterium thrives in low-acid or non-acid foods in the absence of air. Foods in this category include corn, beans, peppers, poultry, fish, and meat.

This is the main reason that these types of foods require the higher temperatures achieved during pressure canning.

Boiling canned low-acid or non-acid foods for 10 to 20 minutes before eating them will destroy potential lingering toxins. This added step will give you one more reason to feel safe about eating home canned low-acid or non-acid foods.

Water Bath Canner

- *Equipment*: A large metal enamel kettle with lid and jar rack with the capacity to hold up to 7, quart-sized Mason jars. Widely available at discount stores and hardware stores. Cost: $20 to $25.
- *Uses*: The water bath canner is used to process Mason jars for home preservation of jams, fruits, fruit juices, and pickled vegetables.
- *Safety Considerations*: It is a large kettle of boiling water so be careful to avoid steam burns and splashing hot water.

Tested recipes are widely available from the USDA, numerous University agriculture departments, Mason jar manufacturers, fruit pectin manufacturers, and from friends and family who have recipes that are known to be safe.

As you familiarize yourself with the principles of safe home canning, you will be able to judge all recipes for safety and even develop your own.

How to Use Water Bath Can

Please always review the manufacturer's directions for the Mason jars that you are using and pay close attention to the head space and processing times specified by recipes.

- Pick a recipe for the produce that you wish to can. Select only quality produce at or near the peak of ripeness. Prepare the recipe.
- While preparing the food, you will also need to get the Mason jars ready. Wash the jars, bands, and lids in hot soapy water. Only use new lids. Never re-use lids. Jars and bands can be re-used.
- Fill the water bath canner and start to heat it in order to sterilize your jars and lids. Use a large thermometer (a candy thermometer works well) to monitor the temperature. When the water is at least 180°F but less than boiling (212°F) add the jars and lids to the water. Sterilizing bands in unnecessary. Kerr and Ball lids specify that they should not be boiled, so make sure that the water stays just below the boiling point.
- Remove the jars and lids from hot water when the food is ready to be packed in jars.

- Add prepared food to jars. Use the handle of a wooden spoon along the insides of the jars to work out any air bubbles. Leave the specified head space, usually ¼ inch or ½ inch.
- Wipe the edges of jar mouths very carefully. They need to be completely clean. Any food particles or other debris on the mouth edge will interfere with the sealing of the lids.
- Place lids on jars and screw on the bands. Only screw them on hand tight. You don't need to twist hard.
- With a jar lifter, place the Mason jars into the hot water. Make sure that jars are not touching each other or the side of the kettle.
- Bring the water to a boil. Once the water is boiling, you can begin the timer for the processing time specified by the recipe. Adjust the heat source as necessary to keep the water boiling but to prevent it from boiling over.
- Once processing time is complete, shut off heat source, and use a jar lifter to remove jars from the canner. Set the jars on a cloth in a location free of drafts. Do not disturb the jars for 12 to 24 hours. You will likely hear the lids "pop" or "snap" down not long after removal, but the jars need to cool completely to make sure the seal is complete. (Processing times change with your land elevation. Consult charts that come with recipes and/or equipment.)

Pressure Canner for Home Canning

- *Equipment*: A pressure canner that has a sealed lid and a gauge that measures the pressure created by boiling water and steam. Under pressure, the steam will achieve temperatures of 240 to 250°F. Such high temperatures are necessary to destroy the bacterium *Clostridium botulinum*. Canners for home use typically cost between $80 and $130. Major brand is Presto.
- *Uses*: To process canning jars of fruit, vegetables, meat, fish, and poultry for purposes of preservation. Pressure canning is the only safe method for preserving low-acid vegetables and meats. Note, the pressure canner can also be used like a pressure cooker to prepare meals.
- *Safety Considerations*: Proper use of a pressure canner requires diligent monitoring of the pressure gauge during operation and maintenance of the plugs, gaskets, and metal parts. An actual explosion of the equipment would only occur if the heat was left on and the pressure climbed into the danger zone.

Even so, the plug in the safety valve should blow out before a catastrophic failure of the pressure canner happened. Note that the danger zone for the equipment is several pounds of pressure beyond the pressures necessary for processing foods. To avoid problems, keep the pressure within the safe zone by watching the gauge and adjusting the heat source. In case of emergency, rapid cooling of the canner can be initiated by running it under cold water.

How to Use Pressure Can

Please follow the manufacturer's directions that accompany your specific pressure canner model. In general:

- The pressure canner will be filled with approximately 3 quarts of water (this amount would vary depending on size of canner).
- The water is brought to a boil and the prepared Mason jars are placed in the canner.
- The lid is placed on the canner, sealed, and locked, but the pressure regulator is *not* put in place yet.
- Once a free flow of steam is initiated through the vent, it will be allowed to vent for 10 minutes. (Time may vary depending on size of canner.) Venting steam exhausts air from the pressure chamber.
- After 10 minutes, the pressure regulator is put over the vent and pressure begins to build inside. When the interior becomes pressurized, the air vent/cover lock will rise and completely seal the chamber. Then pounds of pressure will start to accumulate and register on the gauge.
- Bring the canner to the pressure specified by the canning recipe and then maintain that pressure by adjusting heat source as necessary. You will find that once pressure has been achieved, the stove burner no longer needs to be on a high setting. Monitor the pressure closely to make sure it does not rise too far beyond desired pressure. Also, do not let the pressure fall below the pressure required by the recipe. It is important for the food within to be kept at the necessary pressure/temperature for the required amount of time.
- Once the food has processed at the required amount of time at the necessary pressure, turn off the heat source. Do not rapid cool the canner because this would cause jars within to break. (Processing times change with your land elevation. Consult charts that come with recipes and/or equipment.)
- Allow canner to cool until air vent/cover lock drops on its own. Then remove pressure regulator and allow canner to set for 10 more minutes.
- At this point you may unlock the lid and open it. Be careful of the steam that comes out because it will be scalding hot.
- With a jar lifter, remove Mason jars and set them on a cloth in a location free of drafts.
- Do not disturb jars for 12 to 24 hours after removal. They will be exceedingly hot. The contents will continue to boil after removal from the canner. As the jars cool, the lids will seal.

General Tips for Home Canning

You will have to attend to many details while preserving food at home, but it gets easier as you become familiar with the process. The following tips will help you successfully complete your home canning project with great results: sealing impossible.

- Always use new lids. Packages of just lids are widely available for this purpose. Jars and bands may be re-used.
- Always wipe completely clean the jar mouth edges before applying the lid. Particles of food on the jar mouth could interfere with sealing.
- Follow canning guidelines. Don't compromise, substitute, or declare something good enough. Follow the directions and your food will be safe and good.
- When pickling foods, make sure to use vinegar with 5 per cent acidity. Some vinegars only have 4 per cent acidity, but don't use them.
- Allow jars to cool undisturbed for 12 to 24 hours after removing them from the water bath or pressure canner.
- Remove the jar bands and wipe clean the jars and lids before storing. Little bits of food often leak out during processing and you would not want to store the jars with food bits stuck to them. You can put clean jar bands back on for storage if you want to.
- Home canned products have a shelf life of up to 12 months. Label and date your jars as you produce the food. Do not eat food more than 12 months old.
- Select high quality produce at or near the peak of ripeness. You can cut out small blemishes if necessary, but don't use overly damaged foods.

HOW CANNING PRESERVES FOODS

The high percentage of water in most fresh foods makes them very perishable. They spoil or lose their quality for several reasons:

- Growth of undesirable microorganisms-bacteria, moulds, and yeasts.
- Activity of food enzymes.
- Reactions with oxygen.
- Moisture loss.

Microorganisms live and multiply quickly on the surfaces of fresh food and on the inside of bruised, insect-damaged, and diseased food. Oxygen and enzymes are present throughout fresh food tissues. Proper canning practices include:

- Carefully selecting and washing fresh food.
- Peeling some fresh foods.
- Hot packing many foods.
- Adding acids (lemon juice or vinegar) to some foods.

- Using acceptable jars and self-sealing lids.
- Processing jars in a boiling-water or pressure canner for the correct period of time.

Collectively, these practices remove oxygen; destroy enzymes; prevent the growth of undesirable bacteria, yeasts, and moulds; and help form a high vacuum in jars. Good vacuums form tight seals which keep liquid in and air and microorganisms out.

ENSURING SAFE CANNED FOODS

Growth of the bacterium *Clostridium botulinum* in canned food may cause botulism—a deadly form of food poisoning. These bacteria exist either as spores or as vegetative cells. The spores, which are comparable to plant seeds, can survive harmlessly in soil and water for many years. When ideal conditions exist for growth, the spores produce vegetative cells which multiply rapidly and may produce a deadly toxin within 3 to 4 days of growth in an environment consisting of:

- A moist, low-acid food.
- A temperature between 40° and 120°F.
- Less than 2 per cent oxygen.

Botulinum spores are on most fresh food surfaces. Because they grow only in the absence of air, they are harmless on fresh foods. Most bacteria, yeasts, and moulds are difficult to remove from food surfaces.

Washing fresh food reduces their numbers only slightly. Peeling root crops, underground stem crops, and tomatoes reduces their numbers greatly. Blanching also helps, but the vital controls are the method of canning and making sure the recommended research-based process times found in the USDA's Complete Guide to Home Canning are used.

The processing times in this book ensure destruction of the largest expected number of heat-resistant microorganisms in home-canned foods. Properly sterilized canned food will be free of spoilage if lids seal and jars are stored below 95°F. Storing jars at 50° to 70°F enhances retention of quality.

Food Acidity and Processing Methods

Whether food should be processed in a pressure canner or boiling-water canner to control botulinum bacteria depends on the acidity of the food. Acidity may be natural, as in most fruits, or added, as in pickled food. Low-acid canned foods are not acidic enough to prevent the growth of these bacteria. Acid foods contain enough acid to block their growth, or destroy them more rapidly when heated. The term "pH" is a measure of acidity; the lower its value, the more acid the food. The acidity level in foods can be increased by adding lemon juice, citric acid, or vinegar. Low-acid foods have pH values higher than 4.6. They include red meats, seafood, poultry, milk, and all fresh vegetables except for most tomatoes. Most mixtures of low-acid and acid foods also have pH values above 4.6 unless their recipes include enough lemon juice, citric acid, or vinegar to make them acid foods. Acid foods have a pH of 4.6 or lower. They include fruits, pickles, sauerkraut, jams, jellies, marmalades, and fruit butters. Although tomatoes usually are considered an acid food, some are now known to have pH values slightly above 4.6. Figs also have pH values slightly above 4.6. Therefore, if they are to be canned as acid foods, these products must be acidified to a pH of 4.6 or lower with lemon juice or citric acid. Properly acidified tomatoes and figs are acid foods and can be safely processed in a boiling-water canner. Botulinum spores are very hard to destroy at boiling-water temperatures; the higher the canner temperature, the more easily they are destroyed. Therefore, all low-acid foods should be sterilized at temperatures of 240° to 250°F, attainable with pressure canners operated at 10 to 15 PSIG. PSIG means pounds per square inch of pressure as measured by gauge. The more familiar "PSI" designation is used in (the Complete Guide to Home Canning). At temperatures of 240° to 250°F, the time needed to destroy bacteria in low-acid canned food ranges from 20 to 100 minutes. The exact time depends on the kind of food being canned, the way it is packed into jars, and the size of jars. The time needed to safely process low-acid foods in a boiling-water canner ranges from 7 to 11 hours; the time needed to process acid foods in boiling water varies from 5 to 85 minutes.

Process Adjustments at High Altitudes

Using the process time for canning food at sea level may result in spoilage if you live at altitudes of 1,000 feet or more. Water boils at lower temperatures as altitude increases. Lower boiling temperatures are less effective for killing bacteria. Increasing the process time or canner pressure compensates for lower boiling temperatures. Therefore, when you use the Complete Guide to Home Canning, select the proper processing time or canner pressure for the altitude where you live.

Nutrition Value

Canning is a way of processing food to extend its shelf life. The idea is to make food available and edible long after the processing time. Although canned foods are often assumed to be of low-nutritional value (due to heating processes or the addition of preservatives), some canned foods are nutritionally superior-in some ways-to their natural form. For instance, canned tomatoes have a higher available lycopene content.

Potential hazards

Migration of Can Components

In canning toxicology, *migration* is the movement of substances from the can itself into the contents. Potential toxic substances that can migrate are lead, causing lead poisoning, or bisphenol A, a potential endocrine disruptor that is commonly use to coat the inner surface of cans.

Botulism

Foodborne botulism results from contaminated foodstuffs in which *C. botulinum* spores have been allowed to germinate and produce botulism toxin, and this typically occurs in canned non-acidic food substances. *C. botulinum* prefers low oxygen environments, and can therefore grow in canned foods. Botulism is a rare but serious paralytic illness, leading to paralysis that typically starts with the muscles of the face and then spreads towards the limbs.

In severe forms, it leads to paralysis of the breathing muscles and causes respiratory failure. In view of this life-threatening complication, all suspected cases of botulism are treated as medical emergencies, and public health officials are usually involved to prevent further cases from the same source.

HOME FOOD PRESERVATION

Safe Tomato Tips

Today's tomatoes border the fine line between high and low acid foods. To make sure homecanned tomatoes are safe, it's very important to add acid to tomatoes and tomato products. If acid is not added to each jar of tomatoes or tomato products, bacteria can grow and create an unsafe product. The acid may be added directly to each jar before adding the tomatoes or tomato mixture. A little sugar may be added to offset the taste of the acid, if desired. The sugar will not change the acidity level.

Acid	Effect on Tomatoes or Tomato Mixtures	Amount
Citric Acid	Little change in flavour ¼ teaspoon per pint	½ teaspoon per quart
Bottled	Easy to use. Can substitute	2 tablespoons per quart
Lemon fresh squeezed lemon juice because the acid level is uncertain and there is a chance of contamination	bottled lime juice. Do not use	1 tablespoon per pint Juice
Vinegar (5per cent acidity)	Noticeable change in flavour 2 tablespoons per pint	4 tablespoons per quart

Safe Salsa Advice

Most salsa recipes combine onions, peppers and tomatoes with added acid (lemon juice, lime juice or vinegar). The type and amount of ingredients and preparation methods are important considerations in how salsa is canned. If salsa or other tomato-vegetable mixtures are not canned properly, they can support the growth of a potentially harmful bacterium, *Clostridium botulinum*.

To home-can salsa safely, it is extremely important that you follow a tested and approved recipe for home-canning. Food scientists tested recipes to make sure that the amount of acid combined with other ingredients is safe for home-canning, and that the processing times and methods will destroy harmful microorganisms.

The recipe ingredient amounts and processing methods must be followed to the letter. If you enjoy creating your own salsa recipe or tweaking a recipe, plan to freeze it or store it for several weeks in the refrigerator and eat it fresh.

Home-Canned Salsa Guidelines

- Do not reduce the amount of lemon juice, vinegar or tomatoes.
- Do not add extra peppers, onion or garlic but you can reduce the amount of peppers, onion or garlic.
- You can substitute one type of pepper for another.
- Paste tomatoes contribute to a thicker salsa, while slicing tomatoes yield a thinner salsa. Do not drain or squeeze tomatoes to remove the liquid and juices or you will be removing some of the needed acids.
- Red, yellow or white onions may be substituted for each other.
- The amount of dry spices may be altered or deleted.
- For a stronger cilantro flavour, add fresh cilantro just before serving.
- Never add flour or cornstarch to salsa before canning because heat may not penetrate through the jar to destroy harmful microorganisms. Thicken salsas by adding tomato paste or using a paste tomato.

Freeze Peppers

If you have a lot of peppers, store them unwashed in a perforated plastic bag in the refrigerator up to one week. Before using, rinse with cool water. To freeze peppers, choose crisp, tender green, bright red or bright yellow pods. Wash, cut out stem, cut in half and remove seeds.

Peppers may be cut into ½-inch strips or rings. If the peppers will be used in cooked dishes, water-blanch halves for 3 minutes; strips or rings for 2 minutes. Cool promptly, drain and package, leaving ½-inch head space. Seal, label and freeze. If the peppers will be used in uncooked foods, package them raw, leaving no head space. Seal, label and freeze.

Sweet Corn

Sweet corn is an easy vegetable to freeze and produces a tasty, quality product when frozen properly. Harvested corn on the cob is at its optimum quality for 48 hours. When harvesting, do so early in the morning before the heat of the day. Once husked, the ears are blanched in boiling water for 4 minutes, followed by a complete cooling in ice water for about 8 minutes.

Drain and cut the kernels from the cob. An electric knife is a handy tool for cutting off the kernels. Package the corn in freezer containers, leaving ½-inch headspace. Seal and freeze at 0°F or below for best quality. Blanching, followed by chilling in ice water, are critical processes for producing quality frozen corn. The natural enzymes in corn need to be inactivated by blanching to prevent both loss of colour and nutrients, and flavour and texture changes.

Inadequately chilled corn may become mushy due to overcooking the starch. Can you freeze corn on the cob? Yes, it can be done, but with mixed results. Corn lovers are often disappointed with the sometimes mushy, rubbery texture and the cobby taste. It also takes up more space in the freezer. You can enjoy the great taste of summertime sweet corn all year long by following the simple, basic procedures for freezing vegetables.

Pickling: It's Not Just for Cucumbers

Vegetables from asparagus to zucchini can be home preserved by pickling. The key is to select a recipe from an approved source that is specifically designed for the vegetable you are pickling. Follow the directions carefully for a safe, high quality product.

Begin by selecting tender vegetables and plan to pickle within 24 hours of picking. Wash well and drain. Green beans, carrots, and zucchini are raw packed; whereas asparagus is blanched and beets need to be pre-cooked in their skins for 30 minutes. Cider vinegar has a good flavour and aroma, but may darken white or light-coloured vegetables. White distilled vinegar is the preferred choice for pickling.

Follow the recipe recommendations for the ratio of vinegar to water; it does vary by the vegetable. Pickling or canning salt should be used as other salts contain anti-caking materials that may make the brine cloudy. Use fresh whole spices for the best quality and flavour.

Pickled vegetable recipes are developed for pint or ½-pint canning jars. To insure a safe home canned product, water bath process for 5 minutes to 30 minutes as stated in the recipe.

Sauerkraut

A grilled bratwurst topped with sauerkraut, a summertime treat! Sauerkraut can easily be made at home with its basic ingredients cabbage and salt. To make good kraut, begin by selecting disease-free, firm, mature heads of cabbage and begin cleaning and shredding the cabbage within 24 to 48 hours of harvest.

A kraut cutter is the traditional way to shred the cabbage, but a modern-day food processor works well. Canning or pickling salt draws out the cabbage juice so it can ferment. Be sure to use a tested recipe when making sauerkraut as the proportion of salt to cabbage is key to quality kraut.

Using too little salt softens the cabbage and yields a product lacking flavour. Too much salt delays the natural fermentation process. Sauerkraut tips:

- For every 5 pounds of shredded cabbage, mix in 3 tablespoons of canning salt.
- Select a food-grade plastic pail or an old-fashioned crock in good condition to hold the cabbage. Do not use a metal container.
- Cover the cabbage with a heavy plate, or a salt water brine-filled food grade plastic bag to exclude air.
- Store the container at 68° to 74°F while fermenting. The sauerkraut should be ready in 3 to 4 weeks.
- Sauerkraut may be canned and processed in a boiling water bath or frozen.

Half-gallon Jars

Half-gallon jars are hard to find but they are out there. The only home food preservation foods USDA and manufacturers recommend for half-gallon jars are grape juice or apple juice. Grape juice and apple juice are the only high acid juices safe for these large jars.

Currently, there is no research for safe processing times for other juices or products in half-gallon jars. The density of the product, water content and acid levels are the primary variables that affect safe processing times. At this time, it is not possible to determine a safe formula for other foods in large jars using the boiling water bath canner.

5

Mechanism of Food Fortification

INTRODUCTION

Food fortification continues to be a widely used mechanism in many developed countries. In this context, it is the rapidly changing lifestyles and increasing reliance on more highly processed foods has been used to justify the addition of nutrients to an expanding range of foods in order to ensure nutritional adequacy of the diet. It is important to note that there is no general consensus regarding the extent to which food fortification should be practised and the prevailing attitude towards it varies in the developed world.

France, the Netherlands, Norway and Finland have restrictive legislation regarding the addition of micronutrients to foods. Nutrification has, however, continued in these countries for certain processed foods. In France restoration of vitamins to compensate for processing losses to the extent of 80–200 per cent of naturally occurring levels, is allowed. It is also permitted to add nutrients to foods for special dietary uses.

In the Netherlands, addition of nutrients to certain foods for special dietary uses is allowed as well. Vitamin A fortification of margarine and iodine addition to salt are also practised. In Norway and Finland, special permission for nutrient addition to specific foods may be granted. In his recommendations concerning the micronutrient enrichment of foods in the Netherlands, Anderson stated that the increase in processed ready to use foods had not lowered nutrient intakes in that country and that there was no basis for extending the authorisation for food enrichment. In Germany there is currently an effort to promote the fortification of foods with iodine in order to combat iodine deficiency in that country.

In North America, fortification of foods is viewed positively. Many studies underline the role of such nutrified foods in assuring nutritional adequacy in the North American diet. The expanding range of fortified foods can be justified by the fact that RDA's for many nutrients are commonly not met. In a study carried out on a group of women in New York, it was shown that the mean intakes of Ca, Fe and vitamin D were all below the recommended daily allowance. Other factors can complicate food fortification policy, however, as was shown by the USA National Dairy Council's request for monitoring of Ca fortification on the basis that the required element is readily available in dairy products. In developed countries, regulation of fortification is currently receiving more attention than the technologies involved. This is because there is a legitimate fear of over-fortification as manufacturers seek to use fortification as a marketing tool.

Whatever are the issues that now surround food fortification in developed countries, there can be no question of the role which this practice has played in the past. The virtual elimination of micronutrient deficiencies from these countries has been attributed in large part to this.

Food scientists have risen to the challenges posed by micronutrient fortification. Evidence of their success is clearly demonstrated on the supermarket shelves in most developed countries. Technical problems such as the interaction of the added nutrients with the food carrier matrix due to the pro-oxidant and catalytic properties of many of the essential minerals as well as the susceptibility of many vitamins to destruction by heat, light and oxygen have been overcome for an ever widening range of applications.

IMPORTANCE OF FOOD FORTIFICATION IN DEVELOPING COUNTRIES

The 1992 ICN conference held in Rome emphasised the importance of food based activities in their plan of action geared at addressing the issue of micronutrient malnutrition. Food fortification is one of the relevant modes of action. Advantages of food fortification relative to other modes of intervention have been widely noted and a result of these is that fortification programmes can be implemented and yield results within a short period.

The focus of the international community has so far been on the three most prevalent deficiencies: vitamin A, iodine and iron. There are different considerations involved in the establishment of food fortification programmes in developing countries as opposed to developed ones. There are several steps in the implementation of a food fortification programme in developing countries.

Having identified the need for nutritional intervention on behalf of a population or a sub-group thereof and the required levels of fortification, then a suitable carrier must be identified, appropriate fortificants selected, the technologies used in the fortification process determined and some mechanism put in place to determine whether the nutritional objectives of the programme are being met. Selection of the carrier is a critical step and several required characteristics of the carrier have been noted.

The identified vehicle must be consumed in roughly constant quantities throughout the year by the majority of the population. The food must pass through a centralised point to facilitate a rigidly controlled fortification process.

The addition of fortificants at the required levels must not affect the organoleptic qualities of the food.

It is important to emphasise that micronutrient fortification is not universally applicable. In some cases the high degree of decentralisation of food processing activities makes it impossible to regulate or even establish a viable fortification programme. In cases where subsistence farming is practised, fortification is not likely to be a practicable solution to nutrient deficiencies. It is imperative, before embarking on food fortification activity, to carefully define the context in which it is indicated.

ACCESSORY FOOD ITEMS

Although staple foods are generally used as vehicles in food fortification programmes, at times when none can be identified which has all the required characteristics, it has been necessary to find other options. In outlining fortification policy in the Philippines, Parce states that condiments may be fortified but only with nutrients that are deficient in the diet and provided that such food is an appropriate vehicle for the micronutrient and is widely consumed by the general population or is intended for intervention programmes to address deficiency in a specific target population.

SALT

Salt iodisation began in 1922 in Switzerland and has been implemented in many countries as the major mechanism for eliminating iodine deficiency. Today, IDD remains a problem in many countries. WHO and UNICEF have established the goal of 'Universal Salt Iodisation' to be achieved by the end of 1995. Salt has been favoured as a carrier for iodine due to its wide spread coverage, effectiveness, simple technology involved and low cost. Based on the suitability of salt as a widely used and low cost vehicle, fortification of salt with other nutrients has also been attempted.

According to the Codex standard for food grade salt, use can be made of potassium or sodium iodides and iodates. The iodates have been found to be more stable than the iodides under a wide range of conditions. Stability studies of iodised salt using potassium iodate as the fortificant demonstrated that there was no significant loss of iodine on storage in polyethylene bags for up to two years and that boiling of the salt solutions led to negligible iodine loss.

There have been four major technologies used in the addition of iodine to salt. These are dry mixing, drip feed addition, spray mixing and submersion: the salient features of these procedures have been described by Mannar. A thorough review of salt iodisation technology and quality control has been included in the recent text by Mannar and Dunn.

According to Bauernfeind, an acceptable iron fortificant for salt is one which does not discolour the salt nor impart a flavour or odour and remains stable and bio-available on storage. Ferrous fumarate, ferric orthophosphate and ferrous pyrophosphate have all been recommended by INACG for use with salt. Cook and Reuser report that the use of sodium acid pyrophosphate, ferric orthophosphate and ferric pyrophosphate as unacceptable due to poor bioavailability.

To circumvent this problem, iron absorption enhancers, such as ascorbic acid and sodium acid sulphate, have been used. Another possibility is to use less stable iron compounds in conjunction with a stabiliser. Such combinations include ferrous sulphate/hexametasulphate and ferrous sulphate/sodium acid pyrophosphate. Sodium ferric EDTA provides adequate amounts of iron, but has been found to discolour refined salt.

In the fortification of salt with iron, it must first be ground to a coarse powder to facilitate uniform mixing and distribution of the iron fortificant. Rao reported that a ribbon blender or a screw type conveyor mixer with a suitable feeder could be used. He also reported that packaging in unlaminated high density polythene bags was found to be satisfactory.

The addition of both iodine and iron fortificants to salt can lead to the decomposition of the iodine compound and its subsequent liberation. Nestel reported on the use of a fortificant mixture containing 40 ppm of potassium iodate, and 1000 ppm of iron as ferrous sulphate and 10,000 ppm of a permitted stabiliser which rendered good bioavailability of both iodine and iron after prolonged storage. Cost was identified as a constraint in the use of this fortificant system as it added 50 per cent to the retail price of salt.

Fortification of salt with vitamin A has been attempted under labouratory conditions. The fortificant used was dry vitamin A palmitate type 250 SD protected by a lipid. The fortificant was found to be unstable at moisture contents above 2 per cent, since salt is hygroscopic, packaging material with an adequate moisture barrier must be used. Impurities in the salt were also found to destabilise the vitamin A. The particle size and shape must be such that uniform mixing could be achieved and segregation does not occur on storage.

MONOSODIUM GLUTAMATE

MSG is a condiment which is widely used in many Asian countries. It has GRAS status in Codex Alimentarius as a flavour enhancer. This additive has been thoroughly scrutinised due to the allergic response which it has been alleged to induce in susceptible individuals. Recent studies have indicated that it is safe for the vast majority of people. It is worthy of note that the Joint Expert Committee on Food Additives concluded that MSG should not be used in infant foods even though there was no cause for concern about health risks.

Field studies on MSG fortification with vitamin A have been conducted in the Philippines and Indonesia. In the Philippines dry vitamin A type 250 SD was used. This is vitamin A palmitate stabilised in an acacia-lactose matrix.

The MSG was first ground to 100 mesh to facilitate mixing with the fortificant. Problems were encountered with segregation of fortificant and carrier, loss of vitamin A activity and colour deterioration of the fortified product. Vitamin A acetate 325 L, a granular fortificant stabilised in a gelatin - sucrose matrix, was used in place of the finely powdered vitamin A palmitate 250 SD. Under conditions of high humidity, there was hardening of the gelatin coat with associated loss of vitamin potency.

In the Indonesian study, dry vitamin A palmitate type 250 CWS dispersed in an edible carbohydrate, stabilised with antioxidants and coated with a white protective layer, was used. This product was hot and cold water miscible. This fortificant was demonstrated to retain over half of its potency when stored at 25 °C for 18 months. Under moist dark conditions, half of the activity is retained after 7 months. When subjected to light the fortificant is destabilised more rapidly, the uncoated form of the vitamin is less stable under these conditions. The fortificant mixture was added to MSG at the rate of 0.171 wt. per cent. The vitamin A was aggregated into clusters so as to minimise the problem of segregation during mixing and storage. Segregation of the product was still, however, identified as a problem. Colour deterioration of the fortified product was another constraint to the continued application of this technology.

Iron fortification of MSG has also been attempted using micronised ferric orthophosphate and zinc stearate coated ferrous sulphate. The coated ferrous sulphate had reduced bioavailability relative to the uncoated form, but the fortified product was judged to have acceptable colour, taste, bioavailability and particle size properties. Preliminary investigations indicated that the inclusion of dry stabilised vitamin A (type 250 SD) into the iron fortified mixture might be technically feasible.

SUGAR

Sugar has been found to be a suitable vehicle for nutrients in fortification programmes in Latin America and the Caribbean. In the vitamin A fortification of sugar, vitamin A 250-CWS was proven to be the most effective fortificant. The premix is produced by mixing sugar and fortificant in a revolving drum mixer. Physical separation of the fortificant beadlets from the sugar was prevented by the addition of an edible bonding agent. Stabilisers are added by first dissolving in warmed oil under anaerobic conditions, achieved by bubbling nitrogen through the oil, followed by intermittent addition to the fortified sugar with continuous mixing. Peanut, cotton seed, soya bean and coconut oils have all been used in this procedure. It has been suggested that shark liver oil be used in place of the vegetable oils. The natural vitamin A content of this source reduced the need for vitamin A fortificant by 10-12 per cent. The successful use of shark oil would be expected to require rigorous deodourising and refining treatments and stabilisation of polyunsaturated fatty acids. Satisfactory storage stability of the premix was obtained when packaged in double polyethylene bags with an outer paper layer, and stored in a cool, dry place.

The premix was added to the sugar such as to achieve vitamin A levels of 50 IU per g in the final product. Addition of fortificant can take place in the centrifuge at the end of the washing cycle in the sugar refining process, or along the transporting belts prior to packaging. Retention of vitamin A after 6 months at 25 °C was 92 per cent. Higher losses were recorded with storage at 45 °C, with 76 per cent retention of vitamin activity after 3 months.

Fortification of sugar with iron has also been attempted. Bauernfeind reported promising results using sodium ferric EDTA as the fortificant. Segregation of the fortificant and the carrier was not a problem as the iron compound became stuck to the sugar crystals at moisture contents exceeding 1 per cent. A major problem which exists with iron fortified sugar is that on addition to coffee or tea, there is marked discolouration. This phenomenon is reduced with the use of NaFeEDTA, but it is still evident.

SAUCES

Iron fortification of strongly flavoured or dark coloured food products is often simplified due to the fact mat less care is required in consideration of the avoidance of off-flavours and off-colours. Largely for this reason iron fortification of curry powder demonstrated no technical difficulties. Nestel reported that NaFeEDTA was the fortificant used in the enrichment of curry powder in South Africa.

In the fortification of fish sauces the major problem was the formation of a precipitate on addition of iron. When sodium iron EDTA was the fortificant, however, this phenomenon was greatly reduced. The inclusion of iron in an EDTA complex greatly restricted its availability for interaction with macromolecules and other compounds in foods which could eventually lead to precipitation or other deteriorative reaction.

NUTRIENT ADDITIONS TO FOOD PRODUCTS

The following is not an exhaustive review of all food products which are fortified. Food products have been selected for discussion on the basis of their nutritional impact or their applicability to fortification programmes in developing countries.

RICE AND OTHER WHOLE CEREAL GRAINS

Among the cereals, rice presents unique problems in fortification. These are due to the fact that it is most commonly consumed as a whole grain and also in many countries, extensive washing of the grain prior to cooking is the normal practice. The earliest methods of rice enrichment involved the production of parboiled and converted rice. By this means nutrients from the bran layer were transferred to the starchy endosperm.

The parboiling process involved the soaking of the rough rice, the application of heat followed by drying and milling. It was demonstrated that in this way 50-90 per cent of the thiamine was retained. The process for converted rice developed by Huzenlaub was similar to that for parboiling, but also employed pressure differences to facilitate transfer of nutrients. Acid parboiling, described by Kondo was similar to the parboiling except that it was carried out in the presence of acetic acid.

After parboiling and converting, the next methods of enrichment involved the actual addition of nutrients to the milled products. Techniques used for this have been classified into two main groups 'powder type' and 'grain type' enrichment.

In powder type enrichment, a powdered pre-blended mixture of vitamins and minerals has been added at a rate of 1, 0.5. or 0.25 oz. per 100 lbs of rice (a w/w ratio of 1:1600, 1:3200 or 1:6400). For white parboiled rice, the normal practice has been to add the premix soon after milling as the heat and moisture at the grain surface at this point facilitates adherence of the powder. A major disadvantage of this method of nutrient addition has been that 20-100 per cent of the nutrients are lost on washing. In the USA, rice enriched in this way must bear a label stating 'to retain vitamins do not rinse before or drain after cooking'.

In the second major type of enrichment, a powdered nutrient mixture has been applied to the milled rice grains followed by coating with a water insoluble substance. A fortified rice premix produced in this way has then been added to milled rice at a rate of 0.5 per cent, to yield an enriched product conforming to the required standard of identity. In the United States of America, this standard requires between 2.0–4.0 mg of thiamine, 1.2–2.4 mg of riboflavin, 16–32 mg of niacin or niacinamide and 13–26 mg of iron per 100 lbs. of rice. It may contain 250 -1000 USP units of vitamin D and 500 -1000 mg of calcium.

The requirement regarding riboflavin has been stayed pending final action. The objections regarding the use of this vitamin in the premix have been due to its role in the colour deterioration of the enriched grain. 'Grain type' or 'coated grain' enrichment have also been carried out by spraying the premix solution onto the rice which is contained in a rotary cylinder, followed by hot air drying of the rice, application of a water insoluble sealant, addition of the iron compound and finally a second application of the water insoluble compound.

Water insoluble coatings which have been reported include an ethanol or isopropanol solution of zein, palmitic or stearic acid and abeitic acid. Other coating materials have contained ethyl cellulose. They dissolve at the elevated temperatures employed during cooking. Using this 'coated grain' procedure, Cort et al and Rubin et al. were able to successfully enrich rice with niacin, thiamine, pyridoxine, vitamin A, vitamin E, folic acid, iron, and zinc using expanded versions of the procedures already described. The water insoluble vitamins and minerals were added in different layers with intermittent coatings of shellac. Stability to rinsing treatments were reportedly high, with vitamin losses in the range of 0.2-1.1 per cent.

Ricegrowers Co-operative Ltd. (RCL) in Australia revised the HLR rice fortification procedure for the enrichment of rice with thiamine, niacin and iron. The HLR process involved dissolution of vitamins in dilute sulphuric acid, application of vitamin solution to rice, application of a water insoluble substance, application of ferric pyrophosphate and talc followed by reapplication of a water insoluble layer and iron mixture before screening and packaging.

Due to increased cost of raw materials, low factory output, use of hazardous materials, reported problems with browning of enriched grains, RCL modified the procedure to include the application of the nutrient mixture to the rice followed by drying, screening and packing. The nutrient mixture was composed of ferric pyrophosphate suspended in an acid solution of the vitamins. They reported that the browning found in the HLR premix on storage was due to the formation of ferric sulphates and that this problem was eliminated with the use of an alternative acid. Acid hydrolysis at the surface of the rice grain provided a sufficiently sticky surface to ensure adhesion of the iron compound. Cost of the process was reduced and the factory output was doubled. There was no mention, however, of the stability of premix to nutrient loss through rinsing.

In the Philippines, attempts at fortifying rice by application of vitamin A followed by a water insoluble layer were abandoned since washing losses of between 10–20 per cent were recorded. Rice fortification with iron has also been tried in the Philippines using the coating method. The fortificant used in this programme was anhydrous ferrous sulphate. Some discolouration was found after 20 weeks of storage at room temperature and washing losses were 9 per cent.

Ferric orthophosphate (white iron) is a recommended form of iron for use in the fortification of rice. This iron compound is almost water insoluble and has been preferred for mixing with milled rice due to its white colour. When it is oxidised or contains excessive moisture it may become tan, yellow, purple and or black.

Hurrel reported that the bioavailability of ferric orthophosphate varied widely from batch to batch and was highly negatively correlated with particle size as was also found with elemental iron. This fortificant compound is more expensive than anhydrous ferrous sulphate. According to Hurrel the cost of ferric orthophosphate was about six times that of anhydrous ferrous sulphate, for the same level of total iron.

In Japan, a multinutrient enriched rice has been on the market since 1981. The first step in this procedure was acid parboiling in the presence of thiamine, riboflavin, niacin, pantothenic acid and pyridoxine. The second step involved coating the grain with separate layers of vitamin E, calcium and iron in separate layers and finally a protective coating material. Modified atmosphere packaging, utilising aluminium laminate and carbon dioxide, ensured the stability of vitamin E during storage. The yellow or brown colour of the premix was not a problem in this case as it was used in the fortification of 'brown' rice.

Apart from these well established procedures there have been other innovations regarding alternative methods for rice fortification. Joseph et al. described an enrichment procedure using a premix containing thiamine, riboflavin, niacin and pyridoxine. This procedure involved soaking the milled rice in an acid medium containing the water soluble vitamins followed by the cross linking of starch granules in the enriched grains. The cross linking procedure itself was demonstrated to have caused significant vitamin loss, but the added vitamins were highly cook and wash stable. It is possible that this method could have some utility in the future.

Use of fortified simulated grains has featured prominently in attempts at rice fortification in the developing countries. Murphy et al. described the production of such a fortified rice product for use in the Philippines. The synthetic rice grains were produced by extrusion of rice flour in a pasta machine. The best formulation contained vitamin A stabilised by a mixture of tocopherol, ascorbate and lipids with a low level of unsaturation. Retinyl palmitate stabilised in an acacia matrix, type 250 SD was the fortificant used. Retention of vitamin A after washing was reported to be 100 per cent. Vitamin retention after cooking, however, ranged from 60–94 per cent. The formulations which demonstrated the better storage stability, particularly at high humidity, suffered greater cooking losses. Drawbacks with this technology have been reported to exist with respect to blending with the natural product and in the consistency of the simulated grains after cooking. Field trials in Brazil showed no problems with acceptability of rice enriched in this way.

The enrichment of whole wheat grains with vitamin A has been attempted. The fortificant used was a premix comprised of concentrated vitamin A attached to wheat grains, for mixing at a level of 0.25 per cent with wheat grains. The feasibility of this procedure was not determined. The fortification of whole grain cereals with soluble iron compounds is difficult because they promote oxidation of the lipid component of the grain, thus reducing the shelf life.

CEREALS AND CEREAL BASED PRODUCTS

This is one of the most important areas for consideration of food fortification technologies. Foods from this category form the major component of diets around the world. This is especially true in the case of developing countries where dietary diversity is limited. On average, cereals provide 52 per cent of caloric intake globally. For Africans and Asians they represent 60-75 per cent of the caloric intake; for Latin Americans, 50 per cent; and in the United States of America, 26 per cent. In developing countries 95 per cent of the population consume cereals as a dietary staple which also provide about 47 per cent of the per capita protein intake.

Milling of cereal grains prior to their consumption is a common practice. During the milling process a substantial proportion of the nutrients are lost from the refined product. The fortification/enrichment of cereal grains can therefore be rationalised in more than one way. One valid reason is to restore to refined products, nutrients which have been removed during the milling process. Another reason is to improve the nutrient intake levels of target populations which are at risk of micronutrient deficiency.

FLOURS, CORNMEAL AND BREAD AND PASTA

In the fortification of flour the required nutrient mixture is mixed with an appropriate diluent to produce a premix, which is then accurately metered into the flour. The addition of vitamins B_1 and B_2, niacin, iron and calcium to wheat flour is a common practice in many developed countries. It is technologically feasible to add other vitamins and minerals as well. Cort et al successfully used two premixes to fortify wheat flour; the first comprised vitamin A, pyridoxine, folic acid, tocopherol acetate, thiamine, riboflavin, niacin and iron while the second contained calcium, magnesium and zinc.

The vitamin/iron premix demonstrated excellent stability on storage. The flour enriched with both premixes also demonstrated excellent stability on storage at room temperature. Under conditions of accelerated storage at elevated temperature (45 °C), however, there was substantial loss of vitamin A beyond 4 weeks of storage. Parrish et al. also reported good stability of enriched wheat flour stored at room temperature, but about 50 per cent losses in flour stored at 40 °C for 6 months.

The form of vitamin A most commonly used in the fortification of flour was dry stabilised vitamin A palmitate (type 250-sd) powder form. The water soluble vitamins (thiamin, riboflavin, niacin, pyridoxine, folate and calcium pantothenate) are used in pure crystalline form. The mononitrate salt of thiamine is preferred for this use.

Iron is normally used in its reduced elemental form. In the work of Parrish et al. ferrous sulphate was also used as the source of iron and there was no significant difference demonstrated between the flour enriched with ferrous sulphate and that enriched with elemental iron. It has been recommended that ferrous fumarate be used in the fortification of wheat flour within the P.L. 480 programme, due to its greater bioavailability. When mixing dry fortificants with dry foods, careful selection of the physical characteristics of the fortificant compound is important so as to ensure adequate mixing and to minimise segregation on storage.

Rubin et al. investigated the stability of bread made from flour enriched by the 6 vitamins and 4 minerals described by Cort et al. They found that the inclusion of calcium and magnesium adversely affected vitamin A retention during the baking process. There was also an off-flavour detected in bread which included magnesium after 5 days of storage. Crumb colour and grain were also negatively affected by the nutrient multimixes tested by Rubin et al., with the greater affect occurring with the nutrient mix which included magnesium. Fortification of bread with zinc salts exerted no adverse effect on loaf volume, flavour or any other index of bread quality.

Iodisation of bread has been carried out in the Netherlands and Tasmania by the addition of 2-4 ppm KIO_3 to the bread improver which was already in general use. KIO_3 has been used in bread production in the past, not as a fortificant but as an oxidising agent to improve dough quality.

The fortification of corn meal with 6+4 vitamin/iron mixture was shown to be technically feasible (Rubin et al., 1977; Parrish et al., 1980). Losses of vitamin A on storage at room temperature for up to six months were greater than with wheat flour but still remained below 20 per cent.

In many countries pasta or noodles are commonly eaten and these can therefore be important vehicles for fortification. The manufacture of such products involves the production of a dough which is then extruded and dried. Enrichment can be through the use of enriched flour or alternatively, wet addition of a dispersion of the required vitamins can be carried out at the dough-making stage.

Vitamin losses during production depend largely on the drying conditions employed. Dexter et al. enriched durum wheat flour with a vitamin mixture containing riboflavin, thiamine mononitrate and niacin for the production of spaghetti. They reported that high temperature drying treatments resulted in significant losses of riboflavin, whereas the other vitamins were stable to the processing conditions. In all cases cooking losses, estimated between 40–50 per cent, exceeded those experienced during processing.

BREAKFAST CEREALS

The fortification of ready to eat breakfast cereals is a wide-spread practice. In the United States the general practice has been to provide 25 per cent of the RDA of thiamine, riboflavin, niacin, pyridoxine, folate, ascorbic acid and vitamin A per 1 ounce. 10 or 25 per cent iron, up to 20 per cent RDA for calcium have also been added; vitamin D has also sometimes been included. Surveys have indicated that fortified breakfast cereals play an important role in ensuring nutritional adequacy of the diets of the U.S. population.

The minerals and the more heat stable vitamins like niacin and riboflavin have been added to the basic formula mix prior to processing. The heat labile vitamins such as vitamins A, C and thiamine are usually sprayed onto the cereals after the high temperature processes such as oven or extruder. According to Johnson et al. two important considerations in the design of the spray system are:

- To ensure uniformity in spray coverage.
- To optimise protection of both vitamins A and D.

As the hydrophobic vitamins cannot form true aqueous solutions, they have been sprayed on in the form of an emulsion. The composition of the emulsion has been shown to greatly affect the vitamin stability. The inclusion of 15-20 per cent sucrose into the spraying emulsion stabilised the vitamin A. It acted by forming a protective coating around the vitamin on drying. This fact explained the greater stability of vitamin A included in raisin bran as compared with cornflakes. In the latter case a no-sugar spray was used. Vitamin E can be added to the spray emulsion, it was at least as stable when added under these conditions.

Good storage stability of the fortified breakfast cereals was reported. This was partly due to the fact that packaging materials with appropriate barrier properties have to be selected to ensure the maintenance of equilibrium moisture contents below about 5 per cent. Above this level there was textural deterioration of the product which was perceived as 'staleness'. Lund reported almost 60 per cent loss of vitamin C from raisin bran on storage for a 12 month period. This was ascribed to the fact that the moisture levels in this cereal were maintained at a higher level to prevent hardening of the raisins. Improved procedures for the coating of raisins with a moisture barrier material prior to mixing with the material can minimise this problem. The addition of ascorbic acid was found to stabilise vitamin A to a greater extent than did phenolic antioxidants. The addition of tocopherols did not provide any additional stability, possibly because they were already present at optimal levels.

VITAMIN AND MINERAL FORTIFICANTS

Prudent handling of vitamin and mineral additives in food processing requires a sound understanding of the characteristics of these compounds: their stabilities to various unit operations, solubilities and reactivities with other compounds. Many forms of these nutrients have been developed to render them more suitable for use under a wide range of applications.

PROPERTIES OF MICRONUTRIENT COMPOUNDS

VITAMIN A

In vivo, this vitamin is generally found as the free alcohol or esterified with a fatty acid. The vitamin is available in pure form by chemical synthesis as vitamin A palmitate or the acetate, or recovered from molecularly recovered fish oil. It is a yellowish oily material which may crystallise into needlelike crystals. Provitamins which are then converted to their active form, serve not only as nutrifyng compounds but also as colourants and anti-oxidants. The most common of these is beta-carotene.

Vitamin A is quite stable when heated to moderate temperatures in the absence of oxygen and light. Overall loss of activity during anaerobic heating may range from 5-50 per cent, depending on time, temperature and nature of the carotenoids. In the presence of oxygen and light, there can be extensive loss of vitamin A activity through oxidation. The presence of trace metals accelerates this reaction.

In dehydrated foods, vitamin A and provitamin A are highly susceptible to loss by oxidation. The extent of this loss depends on the severity of the drying process, protection provided by packaging materials and conditions of storage. Vitamin A in pure form is unstable in the presence of mineral acids but stable in the presence of alkali.

Naturally occurring vitamin A is insoluble in water but soluble in oil. In this form the vitamin has limited applicability. Vitamin A fortificants are commercially available in a wide range of forms adapted for use under various conditions. For use in fat or oil based foods such as margarines, oils and dairy products, vitamin A as the acetate or palmitate have been used. They are stabilised with a mixture of phenolic antioxidants or with tocopherols.

For mixing with dry products, a dry form of the fortificant was required with the appropriate size and density. Encapsulation of the vitamin in a more hydrophilic coat is commonly practised in order to achieve a more water dispersable product. Two materials used in encapsulation are gum acacia and gelatin. These dry forms of the vitamin are also stabilised using tocopherols or phenolic antioxidants.

VITAMIN D

The principal forms of the vitamin are D_3 and D_2. They are white, crystalline fat-soluble vitamins, formed by irradiation of the appropriate sterol followed by purification procedures. These compounds are sensitive to oxygen and light, with the D_3 form of the vitamin being slightly more stable. Trace metals such as Cu and Fe act as pro-oxidants.

As with vitamin A, commercially available forms include fat-soluble crystals for use in high fat content foods, and encapsulated, stabilised versions of the fortificant, suitable for use in dry products to be reconstituted with water. As was stated for vitamin A, at the levels of water activity which exist in dehydrated foods, these fat-soluble vitamins are most susceptible to oxidative loss.

VITAMIN E

Vitamin E is a slightly viscous, pale-yellow oily liquid obtained from molecular distillation of by-products from vegetable oil refining or by chemical synthesis. The naturally occurring form of the vitamin is the d-isomer. The synthetic compound is a racemic mixture of the d and 1 isomers. The 1-isomer doesn't have the full biological activity of the d-isomer, but due to the stability of the racemic mixture and the ease of purification, the IU of vitamin E has been defined as 1 mg dl-a tocopheryl acetate.

The free alcohol form of the vitamin is highly unstable to oxidation and is therefore widely used in foods as an antioxidant to stabilise the lip id component of foods. Esterified forms of the vitamin, commonly the acetate, are much more stable. For this reason, fortificants are usually of this form. As with the other fat soluble vitamins, cold water soluble forms have been produced by encapsulation within a suitable matrix.

VITAMINS OF THE B COMPLEX

Vitamin B_1, or thiamine, is a white crystalline solid with a characteristic yeast-like odour and a slight bitter taste. Thiamine is produced by chemical synthesis as the hydrochloride and mononitate salts. The hydrochloride is soluble to the extent of 50 per cent in water as compared with 2.7 per cent for the mononitrate.

Thiamine is one of the most unstable vitamins. Its stability to heat and oxidation is greatest at a pH range of 6 and below. At higher values of pH it becomes increasingly unstable. Thiamine is susceptible to nucleophilic attack, therefore it is degraded by some mineral salts in aqueous foods. Thiamine hydrochloride is the fortificant of choice in cases where dissolution in aqueous media is required. In most other cases the mononitrate is used due to its lower hygroscopicity. Thiamine is also commercially available in a coated form using mono- and di-glycerides of edible fatty acids.

Biotin is a white crystalline powder of low water solubility. It is generally commercially available in diluted form as the physiological requirement for this vitamin is so low. Hoffmann-La Roche sells a 1 per cent mixture of this vitamin with dicalcium phosphate dihydrate. Biotin is fairly stable to heat, air and light.

Vitamin B_2, riboflavin, is of an intense orange colour and low water solubility. A commercially available more water soluble form of this vitamin is the sodium salt of riboflavin 5'-phosphate. Riboflavin is generally stable under most processing conditions, but is unstable in alkaline medium. It is very sensitive to light, particularly in the presence of ascorbic acid.

Pantothenic acid, is a pale yellow, viscous, hygroscopic liquid which is very unstable. The most commonly used commercially available form is calcium pantothenate. This is a slightly hygroscopic white powder with no smell but a slightly bitter taste. Stability of this compound is greatest at pH values between 5 and 7.

Vitamin B_6, pyridoxine, is available commercially as the hydrochloride. Coated forms are also available as with all of the B-vitamins. This vitamin is quite stable to heat and atmospheric oxygen and heat, but degradation is catalysed by metal ions.

Niacin in the form of either nicotinic acid or nicotinamide, can be used in nutrient addition to foods. At very high levels, nicotinic acid has been shown to cause unpleasant side effects such as flushing and 'pins and needles'. This has led to some preference for nicotinamide. Both forms of the vitamin are stable to atmospheric oxygen, heat and light in the dry state as well as in solution.

Cyanocobalamin, the most important compound with vitamin B_{12} activity, is commercially available as a crystalline, dark red, hygroscopic powder. Human requirements for this vitamin are very low and it is commonly sold highly diluted by a carrier. In the preparations sold by Hofmann-La Roche, for instance, it can be purchased diluted with mannitol or a mixture containing modified starches, citrate, citric acid, benzoate, sorbic acid and silicon dioxide.

The selection of preparation depends, of course, on the end use. In solution it is most stable between pH values 4-7.

It is unstable to oxidising and reducing agents and exposure to sunlight, but is fairly stable to heat. Folic acid is a yellow-orange, odourless, tasteless crystalline substance. It id moderately stable to heat and atmospheric oxygen. In neutral solution it is quite stable, but instability increases with a shift in pH in either direction. Folic acid is unstable to heat, light, sunlight, oxidising and reducing agents.

VITAMIN C

Vitamin C or ascorbic acid is an odourless, white, crystalline compound which is stable in its dry form. Due to its high water solubility, losses due to leaching can be a problem in some processing procedures. Ascorbic acid is readily oxidised. In dehydrated citrus juices the degradation is dependant on both temperature and water activity. Other factors as well can influence the degradation behaviour of vitamin C, these include salt and sugar concentration, pH, oxygen, metal catalysts and ratio of ascorbic: dehydroascorbic acid.

Vitamin C addition to foods is commonly practised for reasons other than fortification. Commercially available forms of this vitamin include the free acid and the sodium and calcium salts of these, in powder as well as crystalline or granular form. For mixing with dry products, particle size and density are of course important considerations.

A fat coated form of ascorbic acid is also available for enrichment purposes. Ascorbyl palmitate, is a form of the vitamin used for purposes other than fortification. It is used as an antioxidant in fats and oils and has also emulsifying properties. Other areas of food processing for which vitamin C has application are the prevention of browning in fresh and canned fruit and vegetables, acidification, curing of meat and prevention of haze formation in brewed products.

IRON FORTIFICANTS

Iron compounds used in food fortification are commonly classified according to their solubility. Selection of an appropriate iron fortificant for any given application is based on the following criteria: organoleptic considerations, bioavailability, cost and safety.

The colour of iron compounds is often a critical factor when fortifying light coloured foods. For example white iron, ferric orthophosphate, is often the fortificant of choice in the enrichment of rice.

The use of more soluble iron compounds often leads to the development of off-colours and off-flavours due to reactions with other components of the food material. Infant cereals have been found to turn grey or green on addition of ferrous sulphate. Off-flavours can be the result of lipid oxidation catalysed by iron. The iron compounds themselves may contribute to a metallic flavour. Some of these undesirable interactions with the food matrix can be avoided by coating the fortificant with hydrogenated oils or ethyl cellulose.

Bioavailability of iron compounds is normally stated relative to a ferrous sulphate standard. The highly water soluble iron compounds have superior bioavailability. Bioavailability of the insoluble or very poorly soluble iron compounds can be improved by reducing particle size. Unfortunately this is accompanied by increased reactivity in deteriorative processes.

Sodium iron EDTA is less well absorbed than ferrous sulphate from foods which contain few inhibitors to absorption. In the presence of these inhibitors, however, the EDTA complex is better absorbed. Sodium iron EDTA also participates to a lesser extent in deteriorative reactions. The use of this compound reduces the problem of precipitate formation in foods such as fish sauces and tea.

The use of this compound is not advised in developed countries where the population already receives close to the recommended acceptable daily intake of EDTA.

The problem of low bioavailability of some of the less reactive forms of iron is often circumvented by the use of absorption enhancers added along with the fortificant. Examples of such enhancers are ascorbic acid, sodium acid sulphate and orthophosphoric acid.

IODINE COMPOUNDS

The most commonly used compounds in the iodisation of foods are the iodides and iodates of sodium and potassium. These are the additives allowed by Codex Alimentarius in the iodisation of salt. The iodide compounds are cheaper, more soluble and have a higher iodine content (so that less is needed to achieve the same level of iodisation) than the corresponding iodates. Iodates are more stable under conditions of high moisture, high ambient temperature, sunlight, aeration and the presence of impurities.

The use of iodate is therefore recommended for use in developing countries. Potassium iodide is well suited in cases where the salt is dry, free from impurities and has a slightly alkaline pH. Otherwise the iodide may be oxidised to molecular iodine and lost through evaporation.

If excess water is present the iodide may be separated from the salt in the water film. Loss of iodide can be reduced through the addition of stabilisers such as 0.1 per cent sodium thiosulphate and 0.1 per cent calcium hydroxide combined or 0.04 per cent dextrose and 0.006 per cent sodium bicarbonate. Calcium salts have been used with some report of off-flavour due to the calcium ions. The calcium compound is also much less water soluble than the sodium and potassium compounds and this further limits its applicability.

OTHER MINERAL ADDITIVES

A range of mineral salts are available for fortification with Ca, Mg, P, Zn, Cu and Mn. Prudent selection of mineral compounds is based largely on consideration of mineral reactivity and solubility of the salt. The requirements of the fortificant vary according to the nature of the food product and its end use.

To overcome problems of flavour, texture and colour deterioration due to addition of minerals, some companies have engineered new fortificant preparations which generally involve the use of stabilisers and emulsifiers to maintain the mineral in solution.

ECHNOLOGIES ON FORTIFICATION PROCEDURES

In order to achieve the required level of nutrients in fortified products reaching the consumer, manufacturers have to estimate processing and storage losses and add the necessary excess during production. The introduction of new processes, equipment and packaging materials can affect processing and storage losses and hence fortification procedures.

THERMAL PROCESSES

Thermal treatments in food processing serve multiple functions: destruction of microorganisms, inactivation of enzymes and toxic factors, modification of flavour and texture. In many cases the heat treatment is selected based on consideration of the required lethality of the process so as to render the product safe under stated conditions of storage.

The lethality of a process is a function of both temperature and time of exposure at each temperature. The Arhenius activation energy of spore inactivation is high as compared with that for vitamin destruction. For this reason, at high temperatures the microbiological requirements of the thermal process can be met with relatively low losses of vitamin being incurred. Heat treatments based on the principle of 'High Temperature Short Time' (HTST) or Ultra High Temperature (UHT) are less destructive to the vitamin content and require the addition of lower levels of excess vitamins to account for processing losses.

As the lethality of a process is determined by sum of lethal effects at the 'cold spot' of the product unit (can, pouch, etc.), innovations in equipment design which lead to the faster attainment of thermal equilibrium will also affect the thermal process which must be delivered and hence vitamin loss during processing. The introduction of agitating retorts and retorts which can be used in the sterilisation of materials in flexible packages as well as the development of flame sterilisation processes also can have an affect on fortification procedures.

PACKAGING MATERIALS

The use of retort pouches and other deformable containers in thermal processing influence processing procedures and the extent of thermal degradation of nutrients. The important role of packaging in the application of aseptic processing technologies has important implications for fortification procedures. This is not only in terms of the UHT treatments which can be applied to thin liquid films prior to packaging, but also to the impact on storage losses. The shelf life of aseptically processed foods can exceed 1 year and storage losses over this entire period must be taken into consideration in the calculation of required overages.

During aseptic processes the product is cooled prior to packaging into sterile containers, headspace oxygen and dissolved oxygen levels are therefore higher than for hot-filled products or for traditional thermal processes which involve exhausting or vacuum sealing prior to processing. Problems of vitamin loss in aseptically processed foods have been attributed to dissolved and headspace oxygen, oxygen and light permeability of the packaging material and to commodity-specific reaction which are only dependent on storage temperature.

In aseptically processed milk the dissolved and headspace oxygen lead to rapid initial losses of vitamin C activity. This in turn causes substantial losses in folate as vitamin C plays an important role in the protection of folate. Oamen also reported heavy losses of vitamins B_6, and B_{12} during storage of aseptically processed milk.

Improvements in the barrier properties of plastics and laminates also impact on the potential for nutrient loss on storage of products other than those aseptically processed.

IRRADIATION AND MICROWAVES

Irradiation is used to a limited extent in food processing and as such cannot be expected to impact upon fortification procedures to a large extent. One common use of food irradiation is in the prevention of insect infestation in grains. Losses of vitamin B_1 on irradiation of whole grains are small, but increase in the irradiation of milled grains. Losses of this vitamin can be reduced by exclusion of oxygen during irradiation and storage (Kilcast, 1994).

The increased use of microwave cooking on sous-vide and other cook-chill foods does not have a major effect on vitamin retention.

FUTURE DIRECTIONS OF FORTIFICATION

TECHNICAL PROBLEMS REQUIRING ATTENTION

Loss of vitamin A on storage was found to be significant in a number of fortification processes. The solution to this problem might lie in the design of a new fortificant preparation in which vitamin activity is better protected. This approach has, however, in the past led to new problems regarding cost and bioavailability. Alternatively more studies might be carried out to determine critical factors pertaining to packaging materials or storage conditions which affect vitamin content. Problems persist with iron fortification as well. Discolouration of the food vehicle due to the presence of iron hinders consumer acceptance in many cases. With the existing iron fortificants available commercially, trade-offs with respect to reactivity, bioavailability and cost must be accepted. Sodium ferric EDTA has several useful properties in respect of fortification, but its status as a food additive is only now being considered by Codex Alimentarius Commission for use as a dietary supplement for use in supervised food fortification programmes in populations where iron-deficiency anaemia is endemic. JECFA has requested that additional studies be carried out to assess the site of deposition of iron administered in this form.

Loss of vitamin C during storage, particularly due to the presence of oxygen, is a problem with many fortified foods. Possible solutions to this problem include the development of new, more highly protected vitamin fortificants or investigation of new packaging materials or storage conditions which promote increased vitamin stability.

The use of iodised salt in food processing is another area which requires attention. There are at least two distinct issues to be considered: the first is the role of iodine compounds in deteriorative reactions such as off-flavour development and discolouration, the second is the effect of processing operations on the iodine content of the fortified food. Some relevant data for selected food processes have been reported but further research needs to be done.

If, for example, the iodine fortification of fermented vegetable products was to be considered through the use of iodised salt in their production, it would be essential to know the effect of the iodine compound on product quality as well as the stability of the iodine compound under fermentation conditions.

ANALYTICAL CONSIDERATIONS

Many advances have been made with respect to the analysis of micronutrients in foods. According to Walter problems still remain with the analysis of vitamins which occur in very small concentrations such as vitamins D_3, B_{12}, folate and biotin.

Research is required into the development of appropriate analytical methods for use in the field. Such testing is a required component of quality assurance in food fortification programmes. 'Appropriate' is determined by such factors as availability of equipment and materials for testing, accuracy and precision of the testing method, simplicity and rapidity of the experimental procedure. As analytical methodologies continue to be developed or modified there is a need to verify agreement of results obtained using different procedures. There may also be a need to standardise some aspects of experimental procedures.

HEALTH IMPLICATIONS OF MICRONUTRIENTS

Fortification practices are predicated on our understanding of our physiological requirement for the micronutrients. This understanding has continued to be expanded. The possible preventative effect of antioxidative vitamins on cardiovascular disease and cancer is currently the subject of much investigation. Based on the findings of these investigations, an increase in the recommended daily allowances for these vitamins may be indicated.

Since fortification procedures are based upon delivering a predetermined proportion of the RDA for the nutrients in question, such research will impact upon food fortification. Up to the present time, due to high variability between and within individuals, it has not been possible to establish generally accepted cut-off values for the health protective potential of B-carotene, vitamin E and vitamin C.

RDA's and National Reference Values (NRV) for an expanding number of essential trace minerals are being determined. Continued international discussion and agreement on these is essential.

NUTRITIONAL IMPLICATIONS OF NEW DIETARY TRENDS

Consumer demand for low fat products has given rise to concern that the loss of fat soluble vitamins associated with the removal of fat could lead to inadequate intake of these vitamins among certain sectors of the population.

The finding of the paper prepared by Germany for the Codex Committee on Nutrition and Foods for Special Dietary Uses was that it was premature to determine fortification requirements for lower tat products. Their recommendation was for intensive study of consumer behaviour before other measures be taken.

STANDARDISATION AND REGULATION

A thorough review of existing legislation pertaining to fortified foods needs to be carried out. This should provide the basis for suggested changes where it is found necessary. An example of the effect of legislation on fortification procedures was the use of a mixture of table salt and curing salt in the production of certain iodised sausages in Germany since iodine was allowed in table salt but not curing salt.

The General Agreement on Tariffs and Trade (GATT) Uruguay Round of Multilateral Trade Negotiations have had significant implications for the work of the Codex Alimentarius Commission. The GATT Agreements on the Application of Sanitary and Phytosanitary Measures and on Technical Barriers to Trade emphasize the need for international standards, guidelines and recommendations to facilitate international trade. The Codex Alimentarius Commission needs to work further on the development of standards for fortified foods. Such standards must of course be consistent with the general principles of the Commission regarding fortification. International standards must be sufficiently flexible to allow national governments to meet the specific nutritional needs of target populations while still providing guidance with respect to acceptable ranges of fortification and unambiguous definition of terms.

International standards for fortified foods used in food aid programmes need to be considered. This would facilitate the interchangeable use of foods coming from different sources in meeting the needs of target groups, and could provide the standard which quality assurance programmes should demonstrate all food aid products to meet.

6

Food Production Technology

MECHANIZATION, INCOME, AND EROSION

Mechanization, which often came with the green revolution technologies, was said to be bad because it displaced labour. Again, it often happened that the increase in crops and output per year because of varieties that needed shorter growing seasons and higher yields per crop had labour requirements that more than offset the labour saved by mechanization. In most areas, the extra crop, thereby demanding greater labour, would have been impossible without mechanization and the green revolution technologies. In the 1970s, saving labour by mechanization was always assumed to be a bad thing, though it might also mean more leisure for the farmer. Increasingly, mechanization means that farmers may earn cash income off the farm, affording school fees for children or consumption of products in the market economy.

It was conceded that the demand for agricultural labour was up, but that mechanization, by doing the more menial tasks, displaced women and children from the farm labour force. Many who oppose child labour in factories somehow deem agricultural stoop labour by children in the hot sun to be virtuous.

As incomes go up and child labour is displaced, there is a very good chance that they will now be in school. And where incomes go up, other technologies are also acquired that improve the quality of life and save on child labour. The high-yielding varieties of rice have been basically for production in irrigated fields. The upland varieties have barely changed in yields over the last three decades. This has led to the following argument:

The Green revolution has given rise to greater regional inequality of income. Careful empirical studies of rice cultivation in Asia have found that labourers from the upland regions migrate to work in the paddies, smoothing out the potential regional differences.

There are also arguments seeking a reduction in genetic variety, soil erosion, or groundwater contamination, which are of serious concern but cannot negate the enormous gain in human nutrition or the human catastrophe that would result from any attempt to reverse the green revolution. The diminution of genetic diversity is not a phenomenon of the green revolution per se but of agriculture. The individual farmer will select for yield or other output concerns, not for global genetic diversity. Modern no-tillage (or reduced or minimum tillage) agriculture using pesticides for weed and pest control conserves both soil and water better than its organic competitors.

MODERN FOOD SUPPLY AND SAFETY

Regularized and improved food supply has been a consistent factor throughout history in improving human health and longevity. In more recent times, improved food safety has added another dimension to human health. The evidence on the safety of our modern food supply is quite contrary to the strongly held conventional wisdom. Humans have had to deal with food contamination throughout history. Human foodstuffs have been the carriers of botulism, ergot, and the aflatoxins from the fungus *Aspergillus flavus,* which has caused mass illness, blindness, and largescale death. Even uncontaminated natural foods contain substances that would be considered a threat to human life if they were used as a food additive.

In the United States, during the early twentieth century, contaminated food, milk, and water caused many foodborne infections, including typhoid fever, tuberculosis, botulism, and scarlet fever. Once the sources and characteristics of foodborne diseases were identified, there were technologies that could help to control them before vaccines or antibiotics: "handwashing, sanitation, refrigeration, pasteurization, and pesticide application." Healthier animal care, feeding, and processing also improved food supply safety.

PLANT-PRODUCED CHEMICALS

The buzz word "chemicals," used by some to label and condemn food additives, could appropriately be used to describe what the plants themselves manufacture. Plants of all kinds, and most other life forms, are chemical factories, manufacturing a variety of chemicals, some of which have very active properties. Humans have extracted some of these chemicals for medicinal uses, some as poisons. Given the truism that dose makes the poison, some chemicals from plants that were once used as poisons are now used for medical purposes.

In other words, chemicals from plants are not inherently good or bad; goodness or badness is determined by a variety of factors and the uses to which they are put. One of the reasons for maintaining as much biodiversity as possible is that the earth's plants produce an array of active chemicals, the vast majority of which have not been screened for possible human use.

According to Katharine Milton, plants are unable to run from hungry predators and so have developed a variety of defences to avoid the loss of edible components. "These protections include a vast array of chemicals known as secondary compounds (such as tannins, alkaloids and terpenoids). At best these chemicals taste awful; at worst, they are lethal". Leopold and Ardrey add that these plant secondary substances are "chemicals that do not participate in the basic metabolism of the plant. Among these are many chemicals that serve to repel or discourage the use of the plant by insects, microorganisms, nematodes, grazing animals, and man." These plant toxins apparently are often used to ward off fungi.

"While antifungal chemicals in nightshade were distasteful ... there are probably protective compounds in other wild fruits that prevent rotting without spoiling taste, compounds that should be of particular interest to plant breeders". Many plants produce "phytoestrogens," the so-called estrogen disrupters, which might have the effect of reducing the fertility of their predators, the animals that graze or otherwise consume them. "Essentially, they give their predators an oral contraceptive". Eating fresh, uncooked produce may have many taste and health benefits, but it can also be dangerous unless proper precautions are taken. As epidemiologist Michael T. Osterholmm stated: "The hearthealthy diet may be kind to our cardiovascular system but it is hell on the digestive tract".

Rachel Carson, whose book *Silent Spring* is seen as the inspiration of the modern environmental movement, recognized the role of natural toxins—systemic pesticides—in plants. She maintained that the perception of nature as an "enchanted forest" was a "fairy tale". Even some of the weeds associated with cultivars and often eaten with them are endowed with toxins and other substances deleterious to human health. Garn notes, "Weeds growing along with cultivars often included plants with interesting alkaloids, many of them in the tomato family that includes *Datura, Belladonna,* and the like".

TOXINS IN PLANT EVOLUTION

Through time, plants evolved for their own survival, not to serve human needs. "Plants in nature synthesize toxic chemicals in large amounts, apparently as a primary defence against the hordes of bacterial, fungal, and insects and other animal predators *plants in the human diet are no exception.*"

The evolutionary process involved in domestication changed the survival strategy for these plants, as they now depended upon the continued cultivation to survive.

The toxicity of many plants was reduced compared to their wild progenitors, but in most cases was not completely bred out of plants. In fact, currently there is considerable effort to breed plants with greater disease resistance so as to reduce the need for pesticides, in some instances by increasing the production of these toxins; in other instances, not.

GREEN REVOLUTION PROS AND CONS

It is hard to imagine anyone opposing the green revolution. The increase in yields from the green revolution in rice alone has produced enough to feed one billion people. Growth in yields accounts for 92 per cent of the increase in world cereal production since 1960. The world average for grain yields per hectare rose from 1.1 tons in 1950 to 2.9 tons in 1992. Without the increase in yields, it would have been necessary to bring another 3.6 billion hectares of land under cultivation, which would have almost doubled world cropland from "34 per cent of the earth's surface to 61 per cent".

The green revolution was in many ways a grain revolution— wheat, rice, and maize—but it also was a research revolution that facilitated an extraordinary and sustained general expansion in food production. Bender and Smith note, "Between 1961 and 1994, the number of daily food calories per capita rose from about 1,900 to 2,600 in developing countries, while their populations nearly doubled from 2.2 billion to more than 4.3 billion." Globally in the same period, "average daily per capita food supplies increased more than 20 per cent." The increase in available calories per capita for the developing countries rose 50 per cent from 1948–1952 to 1994–1996. A century-long trend of falling real food prices continued during the period 1950 to 1992 as international food commodity prices dropped 78 per cent in constant 1990 prices.

D. Gale Johnson sums up these achievements in agriculture: "People today have more adequate nutrition than ever before and acquire that nutrition at the lowest cost in all human history." In the 1990s, while population continued to increase, the absolute number of malnourished people in the world declined by forty million, which is a continuation of a trend of the last decades. However, this was considered a "failure" by the UN World Food Programme because it fell short of the reductions that were thought to be possible. To the extent that it is a failure, it is a failure of institutions, not of the technology that is producing the food that could make the programme a success.

So, there are arguments critical of the green revolution. The first argument is that it was a failure. Data such as that just cited above have demolished that argument, although a few still assert it, unaware of the change in the antitechnology party line. Next, we are told that the new varieties may have increased yields, but they led to a worsening of income distribution and the condition of the very poor. As early as the 1970s, studies emerged showing that the poor consumers and farmers benefited proportionately more than other groups as part of the overall improvement in human wellbeing.

Some argued that tenant farmers in the Punjab in the north of India were displaced from their land due to high-yielding varieties of wheat. It turns out that they returned to the land as farm labourers and earned more than they did as tenant farmers. Similarly, in southern India, between 1973 and 1994, the average real income of small farmers increased by 90 per cent, and that of the landless— among the poorest in the farm community—by 125 per cent. Further, caloric intakes for small farmers and the landless rose 58 to 81 per cent, and protein intakes rose 103 to 115 per cent. In addition, greater farm income meant more income and employment in the local villages that supplied the farm with goods and services now affordable to those who worked the land.

The green revolution in Indian agriculture is credited with being a major force in poverty reduction in India. From independence to 1970, the per centage of the population of India living below the absolute poverty line remained relatively steady at 55 per cent. In over two decades since the introduction of the green revolution technologies, the per centage of the population of India living in absolute poverty has fallen to 35 per cent. Throughout most of East and Southeast Asia, the early phases of development involved an emphasis on agriculture and food production.

The green revolution technologies in agriculture have played an important role in the dramatic reductions in poverty rates throughout the region, such as in Indonesia from 58 per cent in 1970 to 8 per cent in 1993, and in China from 270 million people in poverty in 1978 to 65 million in 1999. Yao estimates the reduction in rural poverty in China to be from between 596 to 790 million people (75.5 per cent–100 per cent) in 1978 to 57 to 114 million (6.7 per cent–13.2 per cent) in 1996 (447). By either measurement, the magnitude of the reduction in poverty in China is extraordinary and without precedent in human history.

In the time period 1969–1971 to 1990–1992, global population was increasing about 45 per cent while the absolute number of people suffering from chronic undernourishment in developing nations decreased from 917 million or 35 per cent of their population to 839 million or 21 per cent. From 1990–1992 to 1995–1997, the absolute number of undernourished people in the developing world continued to decline, falling to 790 million. The prevalent high number should not be minimized, nor should we be complacent, but neither should the progress achieved be ignored. For we can finish the job of eliminating undernourishment by recognizing the forces that have brought us this far, thereby gaining some understanding of what still needs to be done. Unfortunately, many wish to deny the progress that has been made, offering antitechnology solutions, which would reverse these gains.

ALTERNATIVE MEDICINE

Some of the most dangerous ingested products legally sold are found in health food stores. Herbal dietary supplements (including exotic oriental herbs) and amino acids pills have been causing health problems for some of their users. A chemical additive may or may not be inactive, but the herbs and amino acids are taken precisely because they are active and will alter a bodily function.

Most of them are untested and unregulated. To the extent that testing is done, there is a critical difference between the approach to herbal medicine and that of pharmaceuticals. Drugs are not permitted to be used until proven to be safe and effective, while the philosophy and practice of alternative medicine has operated on the "presumption that treatments are safe and effective unless proven otherwise". Alternative medicine is big business. One study estimated that in 1990, "Americans poured $13.7 billion into alternative remedies, and made more visits to alternative healers than to primary-care doctors". An editorial in the *New England Journal of Medicine* clearly defines the difficulties with alternative medicine: "What most sets alternative medicine apart is that it has not been scientifically tested and its advocates largely deny the need for such testing".

Lacking is "the marshaling of rigorous evidence of safety and efficacy" necessary for approval of drugs and for publication in the "best peer-reviewed medical journals". One respected study has found that Americans are spending $27 billion a year on alternative treatments, which is close to their outof-pocket expenditures on modern medical treatment.

Some herbs and treatments were found to be beneficial, but some were not. Investors have found the stocks of the firms manufacturing these dietary supplements and alternative medications to be attractive, indicating a belief that consumers will buy them, whether proved to be safe and effective or harmful. Even for some herbs that have been used in low dosage for centuries, there are no studies on the impact of the mega doses that are often taken.

The herbs and alternative medicine advocates have succeeded in blurring the line between food and drugs. Herbal extracts sold as food supplements are being added to packaged, canned, and frozen food products—called functional-foods—with labels that imply health benefits and have sufficient growth potential to have attracted some of the major food producers.

Some of the early advocates of alternative foods and medications, like Rodale Inc. and its publication, *Prevention,* recognize the danger in current patterns of use of herbs and dietary supplements. *Prevention's* national survey on the self-care movement reveals "158 million consumers use dietary supplements for their health and spend $8.5 billion each year." It reported that "widespread use of dietary supplements may cause public health problems". There were risks mixing dietary supplements with prescription drugs and treating oneself without a doctor's supervision. The survey estimated that "11.9 million customers have experienced adverse reactions from using herbal remedies and 6.5 million have had problems of this kind when using specialty supplements".

The problems with various herbs are legion, including lack of long-term studies and of any kind of control of dosage intake. Since many who use alternative medical treatment are also taking prescription drugs under the care of medical doctors, there may be substantial health risks, including death, in combining these with herbs and other alternative medications and stopping the use of some herbs after one had been taking them. Serious problems have emerged from ingesting some of the most popular herbs, such as *Echinacea purpurea,* St. John's Wort, or Ginkgo Biloba. Probably the most serious harm can come from the interaction between herbs and prescription medicines. The increasing provision of these herbs by parents to their children is evident in the statistics of the American Association of Poison Control Centers.

One herbal remedy sometimes called ephedra, sometimes called ma huang, sometimes called pseudoephedrine, contains the chemical ephedrine, "a potent drug whether it is synthetically manufactured or produced naturally from a plant". Dr. Robin R. Caldwell says the drug is "chemically almost identical to amphetamine". Not only is ma huang sold in health food stores, it is a listed ingredient in many products sold at truck stops and elsewhere to help drivers stay awake. Pseudoephedrine is in various cold medicines that carry label warnings not to be used by anyone with high blood pressure, heart trouble, or liver problems.

Ma huang is also used in diet and pep pills. Because ephedra is classified as a dietary supplement under the 1994 Dietary Supplement Health and Education Act, the FDA can restrict its sale "only if it receives well-documented reports of health problems associated with it." Brody notes, "The agency took four years, and more than 100 reports of life-threatening symptoms and 38 deaths, to act against ma huang, a stimulant that can prove disastrous to people with heart problems."

Warnings have also been posted for Chinese herbal products containing aristolochia acid, which has been shown to cause cancer and kidney failure. Nearly one hundred women with kidney damage from Chinese herbs at a Belgian slimming clinic were studied, and several were found "in dire need of kidney transplants" and many of them were developing cancers in the kidney and bladder, years after taking the drug".

Those opposing any regulation of dietary supplements are involved in a contradiction that doesn't seem to bother them. These herbal and other dietary supplements are taken because they have some curative effect from an active ingredient or agent. Stated simply, "if it's strong enough to work, it's strong enough to hurt".

TECHNOLOGY AND LIFE EXPECTANCY

The presumed superiority of all things natural in food and other of life's necessities is the other side of the belief that the food supply in affluent societies is contaminated and is part of the larger, firmly held conviction that modern life is unsafe. One wonders, if contemporary life is so polluted, why are we living so long? Riley has argued that since we live longer and are older, we have more time at risk to ill health.

In fact, morbidity has fallen, along with mortality. We live longer, healthier lives. Among older Americans whom Riley deems to "have more time at risk to ill health," there has been a dramatic decline in chronic disability, according to the National Long Term Care Survey, as reported in a 1997 study by the

National Academy of Sciences. "In the last decade alone, senior citizens have experienced a 25 per cent decline in the number of days of restricted activity due to illness". Improvement in the quality of life continues through the last year of life, even as we live longer.

In the workplace where safety has been largely ignored by environmentalists, the Centers for Disease Control and Prevention has found that "work-related deaths declined from 61 per 100,000 workers in 1913 to four per 100,000 workers in 1997" with a 50 per cent decline in the risk of a fatal or disabling accident from 1970 to 1990. Says Mays, "An estimated 23,000 people died from work-related injuries in 1913 compared with 5,100 in 1997, according to the National Safety Council and the U.S. Bureau of Labour Statistics. During this same period, the work force grew from 38 million to about 130 million."

Contrary to Riley, one might argue that the factors contributing to longer life are part of the same cluster of factors that reduce morbidity, including chronic illness. Fogel and Costa argue cogently, and with massive data, that the epidemiology of chronic disease is not separate from that for contagious disease. Beginning in utero or infancy, inadequate nutrition can lead to a vast array (far too many to list here) of deleterious conditions that make the offspring more susceptible to contagious diseases, to chronic illness later in life, and to shorter life expectancy.

ADVANCES IN CHILDBEARING

Giving birth to a child is dangerous to child and mother. One demographic study for the Plymouth Colony in the seventeenth century concluded that one in every thirty childbirths resulted in maternal mortality, which was about 20 per cent of all female deaths. Add in the 25 per cent estimated death rate for infants and children (for both sexes), and the conclusion is that close to half—about 45 per cent assuming roughly equal infant and child death rates for the sexes—of all females died before completing fertility.

This was in a society where, the author found, the "records suggest a standard of life and health that would compare favourably with that of any preindustrial society today". Edward Shorter, in his controversial writing on the history of women's bodies, argued that virtually all women had some lifelong debilitating health problem resulting from giving birth to children. Given the high rates of maternal mortality, it is not unreasonable to assume that the debilitating injury rate was a multiple of at least five, putting the injury rate close to 100 per cent.

Recent studies would seem to indicate that giving birth to a child, even under favourable conditions, limits a woman's ability to make the required "investments in somatic maintenance" for longevity and thereby shortens her life expectancy. In earlier times, when nutrition levels were very much lower, and the disease experiences (both previous and current infections) of the expectant mother were much higher, it is very likely that there was organ damage to mother and child and that the mother's ability to make the "investments in somatic maintenance" for longevity were greatly diminished.

If having children, even under the most favourable conditions, is harmful to the mother, then an additional benefit of lower infant mortality is that it allows a woman to achieve her desired family size at a much lower health cost to her. This clearly reinforces the thesis of Fogel and Costa on the importance of adequate nutrition and protection from disease for the long-term health of both the mother and the child.

Fogel and Costa call their theory "technophysio evolution." They say, "The theory of technophysio evolution rests on the proposition that during the last 300 years, particularly during the last century, humans have gained an unprecedented degree of control over their environment". This control sets humans not only "apart from all other species, but also from all previous generations of *Homo sapiens.*" They add, "This new degree of control has enabled *Homo sapiens* to increase its average body size by over 50 per cent, to increase its average longevity by more than 100 per cent, and to improve greatly the robustness and capacity of vital organs".

The data on the decline in maternal mortality in developed countries (and to some extent for developing countries) in the twentieth century are every bit as dramatic as the declines in infant mortality and increase in life expectancy during this same time period. The 1998 World Health Organization report confirms that the global trend towards longer and healthier lives continues, though concern is raised for the gap in life expectancy and good health between the rich and the very poor.

WHO is now talking about "health expectancy" being as important an indicator of human well-being as life expectancy. Herman says, "Life expectancy...in rich countries and relatively poor ones, has catapulted to marks unthinkable a century ago. In the industrial world 25 years in life expectancy have been gained in the 20th century, more than twice the gain in all previous human history."

LIFE EXPECTANCY

Many of the news articles on the global increase in life expectancy discuss the problems created by this increase and the social and economic changes necessary to accommodate it. The age at which people retire, the funding of retirement and medical care systems, and the need for education and training for second careers are all issues (or problems), among many others, that will need to be addressed. In many cases they are already issues under serious discussion in economically advanced countries. The character of these systems will differ in the world in 2050, when one-fifth of the global population will be over sixty-five, from those few that existed at the beginning of the twentieth century when the first social security systems began and less than 1 per cent of the world's population was over sixty-five. Though the numbers differ from country to country, the magnitude of the change for most individual countries will be roughly comparable. Technological progress does not offer the promise of a problem-free utopia, but it does offer a never-ending stream of challenges with the ever-present possibility of continuing human betterment.

WATER AND PUBLIC HEALTH

Humanity did not regularly drink hygienically clean water until the advent of purification processes, which included adding chemicals, in the early twentieth century. The late eighteenth and early nineteenth centuries brought developments in basic science, which allowed us to understand respiration and statistics that laid the foundations for quantitative medicine.

Lavoisier's work led to the "secularization and demystification of water by analyzing it and showing that it could be broken down into hydrogen and oxygen". In the middle of the nineteenth century, the physician John Snow identified the water from one well as the source for an outbreak of cholera, followed by Louis Pasteur's recognition of the microbial origin of many diseases, some of which were waterborne. Previously, water could be visually clean or ritually clean, but with this new knowledge and chemical intervention, we could have hygienically clean water. In short, "water became an industrial product".

Chlorination of water in the United States began in the early part of the twentieth century and very quickly "produced dramatic reductions in morbidity and mortality associated with waterborne disease, such as typhoid, cholera, amoebic dysentery, bacterial gastroenteritis, and giardiasis." Chlorinating contaminated drinking water stopped a potential typhoid epidemic in Chicago in 1908. The introduction of drinking water disinfection in the United States is credited with reducing the incidence of cholera by 90 per cent, typhoid and leptospirosis by 80 per cent, and amoebic dysentery by 50 per cent. A chlorinated lime solution was used as an antiseptic in hospitals in the nineteenth century. Prior to chlorination of water, diarrhea and enteritis were the third leading cause of death in the United States.

Despite the fact that millions of lives have been saved by the use of chlorine for disinfection of water, there are some who would ban its further use, even though the evidence for its danger is meagre. Chlorination of water is one of the triumphs of the twentieth century, but many people in the world have yet to experience its benefit. The best estimates are that 80 per cent of all diseases and more than one-third of deaths in developing countries are caused by drinking or cooking with contaminated water. About twenty-five thousand children die each day from waterborne diseases. The concerns expressed about the use of chlorine are for the chlorine compounds that emerge in combination with other chemicals in the water. For any other chemical used instead of chlorine, we would have to be concerned with its toxicity *and* with the toxicity of every possible compound that it might form in the effluent from our water system. The environmentalist precautionary principle would simply rule against any substitute, particularly when set against the nearly century-long safe, effective use of chlorine. Because of U.S. risk assessments claiming that chlorine is a carcinogen, Peru stopped chlorinating its drinking water in 1991. What followed was the "largest outbreak of cholera in recent history, which killed nearly 7,000 people and afflicted over 800,000 more". Regulators, "influenced by the public's present tendency towards chemophobia," at times fail to weigh risks and benefits adequately.

An outbreak of *E. coli* O157:H7 occurred in one community in the United States where the spring water used in the public water system was not chlorinated. In response to the outbreak of *E. coli* O157:H7, the community began chlorinating its water supply. Even with seemingly adequate chlorine levels, an indoor swimming pool complex with "extensive water spray features disseminated pathogenic bioaerosols," causing serious lung infections among its young, healthy lifeguards. Among corrective steps taken was the "provision of free residual chlorine in pool water from chlorine gas". In a study in Britain covering the years 1937 to 1986, "defective chlorination" was blamed in eight out of ten outbreaks of disease from public water supplies and in all thirteen outbreaks in private supplies. Costs of not using a technology are often not considered. The cure for not using a technology is to use more technology and use it more intelligently.

BOTTLED WATER AND TAP WATER

In Houston, people are buying bottled water, which is from the same source—the municipal water supply—that they can draw safely from their own tap more conveniently, at far less cost, and without a plastic bottle that will eventually be discarded and add to environmental pollution. The municipal water supply as the source for the water is clearly printed on the labels of the bottles.

"In fact, 25 per cent of all bottled water is tap water drawn from city water treatment plants, according to Bob Brady of the International Bottled Water Association," though the bottler may do things to the water after it is drawn. An estimate by the American Waterworks Association finds that as many as 50 per cent of bottled water manufacturers in the United States obtain their water from municipal water departments. "This is the stupid thing. They buy water from the municipality for about 40 cents for 1,000 litres, they filter it two or three times, then they sell it at 2,000 times that price".

Much of the bottled water that is not from the local municipal water supply is drawn from aquifers in rural areas. In the United States, where the legal principle of "right of capture" prevails, a major bottling company can come in, drill a few wells, and draw vast quantities of water. In Texas, there have been charges (and denials) that companies are drawing down the aquifers, causing the existing wells of local inhabitants to go dry.

The most important factor behind the extraordinarily rapid growth of the bottled water industry is the public perception that bottled water is cleaner and safer and tastes better than tap water. A study of the bacterial content of bottled water by the University of Mississippi questioned whether it is better than most municipal tap water.

FLUORIDE

In the United States, the average cost of tap water is one cent ($0.01) for five gallons. One very important difference is that tap water is regulated for its safety; bottled water is not. In addition, most bottled water lacks chemical additives such as fluoride (only about twenty brands out of more than five hundred in the United States have it), which has contributed to improved dental hygiene.

Alternative, more expensive fluoride sources are, according to the American Dental Association, "not an effective or prudent public health practice." Even the International Bottled Water Association "recommends customers talk to their dentist or doctor about supplements if they are concerned about fluoride deficiency". Those who are improving their drinking water by using various purifiers and filters may be removing fluoride by using distillers or reverse-osmosis units. Speaking for the American Dental Association, Dr. Michael Easley states that he is "concerned about people who are relying on bottled water. They're not getting enough fluoride and may not realise that they are depriving their children, who will pay the price their entire lives".

Easley points out, "At an average cost of 54 cents per person, per year, the cost of a lifetime of fluoridation for an individual is less than the cost of one small dental filling" and results in "more than $80 in avoided dental-treatment costs" per dollar spent on fluoridation. Recent surveys in the United States indicate that 55 per cent of children between the ages of five and seventeen have no tooth decay in their permanent teeth.

Fluoridation of water is also beneficial for adults and older people. There is less tooth decay and bone loss for older people who live in communities that fluoridate their water. Some firms are marketing bottled water with fluoride added, while one health professional is recommending that parents who continue to use bottled water should consider the need for prescription fluoride as it is unlikely that fluoridated toothpastes will provide enough fluoride to maximize protection. All of which is complicated compared to just drinking tap water.

At the rate of 86.5 gallons of public water per person per day, Americans use tap water for drinking, cooking, washing dishes and clothes, bathing, carrying waste, watering lawns and gardens, and many other uses. With rare exceptions, the water emanating from public water departments is of the highest quality, having been subjected to 150 to 200 scientific quality tests, and would be the envy of our ancestors prior to the twentieth century, or of the hundreds of millions of people in the world today without access to clean water, or any reliable water source, clean or otherwise. Many Americans make a fetish of drinking bottled water, thinking they are getting a hygienically superior.

FOODBORNE DISEASES

Foodborne diseases have always been a problem for humans, even for hunting and gathering peoples, says Cohen. Diseases might be caught from wild animals, including rabies or tularemia, which may have been "a significant cause of sickness and death among American Indian populations who regularly handled game and fur bearing animals." Cohen adds, "Handling wild animals or their remains can also result in infection with such other diseases as toxoplasmosis, hemorrhagic fevers, leptospirosis, brucellosis, anthrax, salmonellosis people can encounter a variety of highly lethal anaerobic bacteria, including the agents of gangrene, botulism, and tetanus, if they expose themselves to the intestines of animals while butchering a kill".

Add in trichina worms (trichinosis) and staphylococcal infections, and it is clear that food harvested in the wild is not necessarily clean, wholesome, and natural. Drinking water in the wild can cause "beaver belly" (giardiasis) or be a source of "microorganisms derived from moose, ducks, and geese as well". Water in the wild today, including snow collected from the ground and melted, can be a source of a variety of parasites, including giardia and cryptosporidium. If we think that we eat contaminated food today, and that we ate clean food in earlier times, think again. Some foods require cooking or other forms of preparation to be digestible by humans, such as soaking in water, having husks or shells or other parts removed, or milling and reducing particle size to facilitate digestion.

Even today, with small-scale technologies, grain milling is incomplete, leaving broken pieces of husk, causing stomach lesions and digestive problems. In the past (and in many less-developed areas today), "harvesting and milling techniques resulted in a considerable accretion of grit and little stones, as we see in the worn teeth of our colonial ancestors, Bronze-age and Ironage Europeans, and American Indians." With storage, "weevils were often a regular part of the daily bread". On the bright side, food contaminants, such as weevils, were sometimes important sources of protein in otherwise protein-deficient diets.

HEALTH FOOD FADS

Health food faddists who look with horror on the minutest trace of any chemical (read manufactured chemical) nevertheless will ingest mega doses of natural products (such as herbs) that contain substances known to be toxins or untested compounds or have active products such as amino acids that have not been tested for safety at the dosage in which they are being taken.

To call free-standing amino acids "natural" is a travesty, since they are not naturally present in food in significant quantity nor can they be "extracted from food without first chemically breaking food protein in the labouratory into its constituent amino acids". It is ludicrous to label a bottle of amino acids "natural," and it shows how meaningless the term has become and how useless it is for formulating public policy.

The amino acid pills are a synthetic product, produced in large vats in chemical factories using microbial fermentation, with a feedstock for the microorganism that is so replete with contaminants that it has to be further processed through a carbon filter and by reverse osmosis to produce what rarely occurs in nature, a free-standing amino acid. We get almost all of our amino acids by ingesting complex proteins.

In 1989, L-tryptophan amino acid pills were identified as a cause of a rare muscle disorder, eosinophilia-myalgia syndrome, which eventually resulted in thirty-seven deaths and at least, 1,500 illnesses. Activists blamed it on the fact that the manufacturer used a bacteria, *Bacillus myloliquefaciens,* that had been genetically engineered, and they continue to argue this to the present, despite the fact that the overwhelming evidence is that it was a contaminant resulting from a reduction in the carbon filtration and modification of the reverse osmosis process.

The tragic incident does not reflect well on the testing procedures (or lack of them) by the firms that bought the bulk powder for tryptophan (or other pills) and converted it into "all natural" pills. Conveniently ignored by the activists is their long history of opposing any regulation or even monitoring of these so-called dietary supplements and their ability to pass legislation, before and after the tragedy, which specifically exempts them from precisely the oversight that might have prevented it.

In fact, in 1972 the U.S. Food and Drug Administration tried to withdraw the GRAS designation for amino acids in "dietary supplements because FDA felt that the available data did not allow" the FDA to deem them to be safe. The FDA had a number of concerns, including possible toxic or other adverse effects. Because of an error in its original GRAS designation, the FDA lost a court case that would have allowed it to have some reasonable oversight.

VITAMINS

Vitamins are natural, even if manufactured, as long as they are labeled organic. They too have not been tested for safety at the mega doses that some faddists use them, though clinical studies have shown that they can be helpful when used intelligently. Obstetricians who frequently prescribe multiple vitamin pills for pregnant women and programmes for vitamin A intervention have saved children's eyesight and lives in poor countries. The faddists who would regulate agricultural chemicals out of existence oppose even simple truth-in-labeling requirements for their beloved natural products.

Groups such as the International Advocates for Health Freedom, the Life Extension Foundation, and "a growing number of allies in the 'patriotic' movement" see a secret conspiracy of "international pharmaceutical empires working hand-in-hand with the UN and the FDA to restrict access to lowcost herbs, vitamins and minerals around the world".

SPROUTS

Alfalfa sprouts have been found to be contaminated with *E. coli* O157:H7 and salmonella. Obviously, not all alfalfa and other sprouts are contaminated but the rate of contamination for raw, untreated sprouts is considerably higher than for most other food. The Emerging Infections Programme of the Centers for Disease Control has estimated that twenty thousand people in North America contracted salmonella from alfalfa spouts in 1995. Disease outbreaks have led to steps such as irradiating alfalfa seeds. Irradiation of alfalfa sprouts is currently being evaluated as an "adjunct seed decontamination method"

Raw alfalfa and clover sprouts are recognized sources of foodborne illness in the United States. Raw sprouts present unique food safety problems because the warm humid conditions needed to grow spouts are also ideal for the rapid growth of harmful bacteria. The very natural nature of alfalfa sprout production creates the contamination problem and makes them a "well-suited vehicle for salmonellosis." *JAMA* points out, "Seeds are often stored for months or years under cool, dry conditions in which salmonellae are stable. During the three- to five-day sprouting process, numbers of salmonellae may increase by three to four orders of magnitude, decreasing little if at all during subsequent refrigeration."

In addition, the way that alfalfa sprouts are stored, distributed, and marketed poses a danger to the health of the user:

From farm to table, many opportunities exist for contamination of alfalfa seeds or sprouts. Crops can be easily contaminated with dirty water, run-off from adjacent farms, animal fertilizers used in previous growing seasons, or droppings from rodents or ruminants.

The natural way that alfalfa sprouts are consumed at home or even in restaurants also adds to the danger. "Alfalfa sprouts are rarely washed or cooked before consumption, and consumers are left with little protection other than chance". The salmonella organism is believed to be present in seed crevices between cotyledon and testa and therefore is not amenable to chemical treatment. At present there is a "fundamental problem" in that the commercial "sprouting process contains no 'kill step' that would eliminate pathogens without compromising a seed's germination potential. Consequently, because of the difficulties in producing a "pathogen-free seed," warnings have been issued about the dangers of eating alfalfa and other sprouts.

The authors of the study conclude: "Alfalfa sprouts are a highrisk food for salmonellosis. All consumers, particularly those at greatest risk for severe disease (immunosuppressed, elderly and very young people), should consider this danger when deciding whether to eat alfalfa sprouts".

Proctor adds, "alfalfa sprouts contain 1.5 per cent by dry weight of canavine, a highly toxic substance". Not only is natural not necessarily better, food processing is an essential component of food safety. Says Garn, "Some cultivars are quite toxic, unless properly prepared".

PRODUCT CHARACTERIZATION

Product characterization allows for consideration of any potential hazard(s) that might be associated with anticipated exposure to a plant-incorporated protectant. This process includes a review of the source of introduced genetic material, including all regulatory sequences, the modifications that have been made, methods of transformation, inheritance, stability and expression in the host plant. For the introduced pesticidal substance, the anticipated modes of action, specificity and toxicity to both target and non-target organisms are evaluated. Target organisms may include weeds, insects and diseases (viral, bacterial, and fungal) affecting a crop. The crop itself, as a volunteer weed, may also be of concern following crop rotations. Once a PIP and any potential for exposure have been characterized, the data required in support of an environmental and human health risk assessment can be defined.

ECOLOGICAL EFFECTS

Environmental fate and exposure are important considerations when evaluating the potential impact of PIPS on the environment or on non-target organisms (*e.g.* wildlife, beneficial insects). Depending on the crop and introduced pesticide substance, data on the level of expression and rate of degradation may be required, first for the tissues of the modified plant, and secondly, the fate in soil, water, or any other environmental compartment.

Such data allow the potential for exposure to the pesticide substance to be evaluated for non-target organisms via feeding on, or through decomposition of, the modified plants.

Studies are arranged in tiers, starting with Tier I, consisting of acute toxicology and basic characterization studies that help predict environmental fate. Depending on the results obtained in Tier I, and the level of concern, additional studies may be requested from higher tiers. It is expected that most PIPS will only undergo Tier I testing, since negative results are expected at relatively high exposure levels.

The 'maximum hazard approach' when used in Tier I provides a high level of assurance that adverse effects would not be expected to occur under actual conditions of use within the field. If the results of Tier I reveal a potential hazard at doses within the range of expected concentrations in the environment, additional studies would be requested in higher tiers to better characterize the potential for any kind of effect. Data requirements may be satisfied by generating test data or by requests for data waivers with credible justification.

The type and extent of dietary studies required will depend on properties of the crop and on the introduced pesticide substance. Dietary studies typically involve feeding plant tissues (*e.g.* grain, pollen) that express the PIP or that have been spiked with the pesticidal substance. However, when determined necessary, the pure pesticidal substance may be administered to non-target species at doses of up to 10 to 100 times the expected level of exposure in the field. Dependent upon the potential for exposure, additional studies may also be requested involving any number of other non-target organisms (*e.g.* honeybees, green lacewing, ladybird beetles, parasitic wasps, earthworms, etc.). When toxicity is observed, the potential for exposure becomes an important consideration when ascertaining whether an adverse effect might be expected under actual conditions of use in the field. The requirement for provision of a resistance management plan is not routine practise for PIPS, but depends on an evaluation of the potential for resistance to develop and the threat to existing or functionally equivalent pest management practises.

To date, resistance management plans have only been placed on Bt PIPS. If considered appropriate, the EPA may request that a resistance management plan be submitted as a part of a condition for registration. A resistance management plan may require a commitment towards the generation of additional research data, to implement structured refuges, to perform annual resistance monitoring, to initiate remedial action plans, and to educate growers. The EPA has worked closely with academia, other federal agencies, public interest groups, industry, and growers, to establish resistance management plans based on the most current science.

HUMAN HEALTH EFFECTS

For agricultural food crops, special consideration is given to evaluating the safety of the pesticidal substance introduced as a component of consumed foods. Rodents (often mice, to conserve test material) are orally administered the pesticide substance at doses of up to 100 000 times the levels of expected human consumption as a part of the diet.

Although the modified plant tissues would be considered the most appropriate vehicle for determining the acute toxicity of a PIP, it is not typically feasible, since only small amounts of the pesticidal substance are present in plant materials.

Therefore, an alternative source (*e.g.* microbial) is often used to generate a sufficient amount of identical pesticide substance for testing purposes. In this case, the equivalency of the test substance obtained from an alternative source to the PIP (*e.g.* composition, identity, etc.) being expressed within the plant is an important consideration that must be well documented.

For pesticide substances that are proteins, the principal concern relates to the potential for the introduction of a new allergen or toxin into the food supply. Since most allergenic food proteins are stable to digestion, in vitro digestibility studies are used to predict how long a pesticidal substance may persist in digestive fluids (*e.g.* gastric, intestinal) following ingestion. In addition, heat stability studies are performed to provide an indication of the fate of the PIP during food processing. Commonly, assays of amino acid sequence homology are performed to determine similarity to substances that are known to pose unique safety concerns (*e.g.* toxins, allergens).

The EPA uses a 'decision tree' approach similar to that followed by the FDA in assessing the potential allergenicity of a new food protein. In many instances, US EPA and US FDA collabourate with each other when evaluating issues of safety for new pesticidal proteins being expressed in GM food crops.

Residue Tolerances for Plant-incorporated Protectants

Analogous to the consideration of tolerances for residues of conventional pesticides, the US EPA reviews data on the human, animal and environmental safety of PIPS to determine whether residue tolerances should be established for the amounts expected to be present in foods derived from the GM plant. Under Section 408 of the FFDCA, if the GM plant expressing a PIP is a food crop, the US EPA is obligated to establish the 'safe level' of pesticide residue allowed, or tolerance level.

The Food Quality Protection Act (FQPA) served to amend the FFDCA and FIFRA by outlining the process by which the safety of pesticide residues on raw or processed foods can be established. Under this process, in order for a pesticide residue to be considered safe, a 'reasonable certainty of no harm' must exist assuming aggregate exposure of the public, taking into consideration sensitive children within the population.

To date, all pesticidal proteins that have been registered as PIPS have been determined to be non-toxic, and as a result, have been considered exempt from the requirement to establish a tolerance under the FFDCA. Consistent with categorical exemptions as originally proposed for plant-pesticides under FIFRA, the US EPA also proposed that tolerances might not be required for pesticidal substances derived from sexually compatible plants, regardless of the process used to introduce the PIP, provided that the genetic material encoding the pesticidal substance was from a plant commonly used as food, and that the presence of the pesticidal substance in a host plant did not result in any new or significantly different dietary exposures. Final rules include exemptions from tolerances for the residues of PIPS, and their encoding nucleic acids, but only for PIPS derived through the conventional breeding of sexually compatible plants. US EPA is requesting further comment on the possible exemption from the requirement of a tolerance for:

- PIPS derived through rDNA technology;
- PIPS that act by affecting plant defences;
- PIPS developed based on viral coat proteins.

Future Considerations

The US EPA anticipates that the requirements for the registration of PIPS will continue to evolve as the science and policies relating to biotechnology continue to mature. In light of on-going criticisms that the plant-pesticide rule has received since its publication in 1994, a number of stakeholder workshops have been hosted by the US EPA. In March 1999, the House Agricultural Subcommittee hosted a congressional hearing to address a number of these concerns. One of the primary questions addressed was whether the definition of a pesticide under FIFRA provides the US EPA with the authority to regulate pest-protected plants that have been developed using rDNA technology. Another concern was that the proposed rule was too broad in scope because all plants have natural defence mechanisms, and as such, would be classified as 'plant-pesticides'.

In addition, an exemption from the plant-pesticide rule for those plants not developed using rDNA technology would serve to regulate the process used in the development of the GM plant, rather than on the pesticidal substance being produced within the plant. Further criticism related to the use of the term 'plant-pesticide' in relation to the products of the genes that had been introduced into GM crops.

As a part of its final rule, the US EPA recommended that plant-pesticides be renamed PIPS, an alternative name which distinguishes them from the other types of pesticide products requiring registration with the US EPA. A new FIFRA section has been added to the Code of Federal Regulations specific to the regulation of PIPS. In doing so, the US EPA has acknowledged that the new final rule would be making a distinction between the products of conventional breeding and rDNA technology, *i.e.* regulating on the basis of process not product.

One reason given for this divergence was that the use of this criterion would be expected to provide the public with increased confidence that an appropriate level of regulatory oversight was in place for assessing the risks of rDNA technology, which is a societal not scientific issue. SAPs are charged with providing the Office of Pesticide Programmes (OPP) of the US EPA with advice and guidance on human health and environmental issues relating to proposed regulatory decisions for pesticides, including PIPS, based on the latest scientific data. For example, the SAP for Bt Plant Pesticides met three times in the year 2000 to consider issues related to the safety of plants that had been genetically modified to express insect-specific Bt toxins. The US EPA released a decision to temporarily extend registrations for all currently registered pest protected plants throughout the 2001 growing season.

Additional provisions, such as the recently strengthened resistance management requirements for plant-pesticides, along with the original registration conditions, were considered to provide adequate protection during the extended time period. In addition, the reassessment process for these products has been designed to assure maximum transparency and opportunity for public comment in the regulatory decision-making process.

EFFECTS ON FOOD PRODUCTION

Assuming no effects of climate change on crop yields and current trends in economic and population growth rates, world cereal production5 is estimated at 3286 million metric tons (mmt) in 2060 (cf. 1795 mmt in 1990). Per capita cereal production in developed countries increases from 690 kg/cap in 1980 to 984 kg/cap in 2060. In developing countries (excluding China) cereal production increases from 179 to 282 kg/cap. Aggregated world per capita cereal production grows from 327 kg/cap in 1980 to 319 kg/cap in 2060. The declining aggregate trend for the future is caused by the relatively large difference in per capita cereal production in the developed and developing countries and the demographic changes assumed by the model.

Cereal prices are estimated at an index of 121 (1970 = 100) for the year 2060, reversing the trend of falling real cereal prices over the last 100 years. This occurs because the BLS standard reference scenario has two phases of price development. During 1980 to 2020, while trade barriers and protection are still in place but are being reduced, there are increases in relative prices; price decreases follow when trade barriers are removed. The number of hungry people is estimated at about 640 million or about 6 per cent of total population in 2060 (cf. 530m in 1990, about 10 per cent of total current population).

ADJUSTMENTS IN THE ECONOMIC SYSTEM

The BLS includes the ability to simulate adjustments that the world food system might make to changes of yield (*e.g.*. reallocation of agricultural land use, change in fertilizer use, and application of irrigation water). Simulations of the effects of climate change without such internal adjustments are of theoretical interest only as these would unrealistically imply no economic or behavioural response of producers and consumers. However, as a measure of distortion of the economic system these hypothetical impacts help to define the adjustments taking place in the system over time. Under these conditions the effects of climate change and increased atmospheric CO2 on crop yields derived from the GCM scenarios imply a 5 per cent to almost 20 per cent reduction in total cereal production. These estimates are changes to production levels projected for 2060 without climate change. Adjustments within the economic system tend to counteract negative yield impacts as agricultural production shifts to regions of more favourable comparative advantage. The BLS offsets 65-80 per cent of the potential impact on yield in scenarios for impacts below 10 per cent of global cereal production (the GISS and GFDL climate change scenarios). The offset decreases to 60 per cent under a scenario of greater yield reduction (*e.g.*., UKMO).

Adjustment in the Economic System

Changes in cereal production, cereal prices, and people at risk of hunger estimated for the GCM doubled CO2 climate change scenarios (with the direct CO2 effects taken into account). These estimations are based upon dynamic simulations by the BLS that allow the world food system to respond to climate-induced supply shortfalls of cereals and consequently higher commodity prices through increases in production factors (cultivated land, labour, and capital) and inputs such as fertilizer. The testing of climate change impacts without farm-level adaptation is unrealistic, but is done for the purpose of establishing a baseline with which to compare the effects of farmer response. We can safely assume that at least some farm-level adaptations will be adopted, especially techniques similar to those tested in Adaptation Level 1 that do not imply major changes to current agricultural systems.

Under the GISS scenario (which provides lower temperature increases) cereal production is estimated to decrease by just over 1 per cent, while under the UKMO scenario (with the highest temperature increases) global production is estimated to decrease by more than 7 per cent. The largest negative changes occur in developing regions, which average-9 per cent to-11 per cent, though the extent of decreased production varies greatly by country depending on the projected climate. By contrast, in developed countries production is estimated to increase under all but the UKMO scenario (+11 per cent to-3 per cent). Thus, disparities in crop production between developed and developing countries are estimated to increase.

Price increases resulting from climate-induced decreases in yield are estimated to range between 25 per cent and 150 per cent. In the case of the GISS scenario, the 5.3 per cent reduction in yields of the unadjusted scenario causes a disequilibrium that is resolved via market mechanisms in the adjusted case. This results in a-1.2 per cent consumer response and about a +4 per cent (relative) producer response and leads to 24 per cent higher relative prices for cereals. While this price response seems to be high, cereal prices only account for a modest fraction, perhaps one third or less, of retail food prices.

Hence, a 24 per cent increase in world cereal prices does not imply a 24 per cent increase in food prices.

These increases in price are likely to affect the number of people with insufficient resources to purchase adequate amounts of food. The estimated number of hungry people increases approximately 1 per cent for each 2-2.5 per cent increase in prices (depending on climate change scenario). People at risk of hunger increase by 10 per cent to almost 60 per cent in the climate change scenarios tested, resulting in an estimated increase of between 60 and 350 million people in this condition (above the reference case of 640 million) by 2060.

Farmer Adaptation

Globally, both minor and major levels of adaptation help restore world production levels, compared to the climate change scenarios with no adaptation. Averaged global cereal production decreases by up to about 160 mmt (0 per cent to-5 per cent) from the reference case of 3286 mmt with Level 1 adaptations. These involve shifts in farm activities that are not very disruptive to regional agricultural systems. With adaptations implying major changes, global cereal production responses range from an additional 30 mmt to slight increase to a slight decrease of about 80 mmt (+1 per cent to-2.5 per cent).

Level 1 adaptation largely offsets the negative climate change yield effects in developed countries, improving their comparative advantage in world markets. In these regions cereal production increases by 4 per cent to 14 per cent over the reference case. However, developing countries are estimated to benefit little from this level of adaptation (-9 per cent to-12 per cent change in cereal production). More extensive adaptation (Level 2) virtually eliminates global negative cereal yield impacts derived under the GISS and GFDL climate scenarios and reduces impacts under the UKMO scenario to one third.

The effects of climate change, and climate change with both levels of adaptation, on cereal prices in 2060. As a consequence of climate change, world cereal prices are estimated to increase by about 25 per cent to almost 150 per cent. Under Adaptation Level 1, price increases range from 10 per cent to 100 per cent; under Adaptation Level 2, cereal price responses range from a decline of about 5 per cent to an increase of 35 per cent. As a consequence of climate change and Adaptation Level 1, the number of people at risk from hunger increases by about 40 million to 300 million (6 per cent to 50 per cent) from the reference case of 641 million. With more significant farmer adaptation (Level 2), the number of people at risk from hunger is altered by between-12 million for the GISS scenario and 120 million for the UKMO scenario (-2 per cent and +20 per cent). These results indicate that, except for the GISS scenario under Adaptation Level 2, the simulated farm-level adaptations did not entirely mitigate the negative effects of climate change on potential risk of hunger, even when economic adjustments, *i.e.*. the production and price responses of the world food system, are taken into account. For each of these alternate future assumptions, a new reference scenario was established with the BLS, and then tested with the GCM climate change scenarios.

Full trade liberalization: Assuming full agricultural trade liberalization and no climate change by 2020 provides for more efficient resource use. This leads to a 3.2 per cent higher value added in agriculture globally and 5.2 per cent higher agricultural GDP in developing countries (excluding China) by 2060 compared to the original reference scenario. This policy change results in almost 20 per cent fewer people at risk from hunger. Global cereal production increases by 70 mmt, with most of the production increases occurring in developing countries.

Climate change impacts were then simulated under these new reference conditions. Under the same trade liberalization policies, global impacts due to climate change are slightly reduced, with enhanced gains in production accruing to developed countries. Losses in production are greater in developing countries. Price increases are reduced slightly from what would occur without full trade liberalization, and the number of people at risk from hunger is reduced by about 100 million.

Reduced rate of economic growth: Estimates were also made of impacts under a lower economic growth scenario (10 per cent lower than reference). Lower economic growth results in a tighter supply situation, higher prices, and more people below the hunger threshold.

The effect of climate change on these trends is generally to reduce production, increase prices, and increase the number of people at risk from hunger. Developed countries increase cereal production in the GISS and GFDL scenarios even with the projected lower economic growth rates, but developing countries decrease production under all climate change scenarios.

Altered rates of population growth: Lower population growth has a significant effect on cereal production, food prices, and number of hungry people. Simulations based on rates of population growth according to UN Low Estimates result in a world population about 17 per cent lower in year 2060 when compared to UN Mid Estimates as used in the reference run. The corresponding reduction in the developing countries (excluding China) would be about 19.5 per cent, from 7.3 billion to 5.9 billion. The combination of higher GDP/capita (about 10 per cent) and lower world population produces an estimated 40 per cent fewer people in hunger in the year 2060 compared to the reference scenario.

Even under the most adverse of the three climate scenarios (UKMO) the estimated number of hungry is some 10 per cent lower than the estimated reference scenario without any climate change. Increases in world prices in agricultural products, in particular of cereals, under the climate change scenarios employing the low population projection are around 75 per cent of those using the UN Mid Estimate. The generalized relative effects of different policies regarding trade liberalization, economic growth and population growth on the production of cereals and people at risk of hunger. Alternative development assumptions make little difference with respect to the geopolitical patterns of the relative effects of climate change. In all cases, cereal production decreases, particularly in the developing world, while prices and population at risk from hunger increase due to climate change.

The beneficial effects of trade liberalization and low population growth are of the same or even greater (in the case of population) order of magnitude as the adverse effects of climate change. This suggests that there may be much to be gained from altering the conditions of trade and development as a strategy for addressing the climate change issue. The magnitude of adverse climate impacts are least, however, under the conditions of low population growth. An assumption of low population growth rate minimizes the population at risk from hunger both in the presence and absence of climate change in the BLS simulations.

Climate change induced by increasing greenhouse gases is likely to affect crop yields differently from region to region across the globe. Under the climate change scenarios adopted in this study, the effects on crop yields in mid-and high-latitude regions appear to be less adverse than those in low-latitude regions.

However, the more favourable effects on yield in temperate regions depend to a large extent on full realization of the potentially beneficial direct effects of CO2 on crop growth. Decreases in potential crop yields are likely to be caused by shortening of the crop growing period, decrease in water availability due to higher rates of evapotranspiration, and poor vernalization of temperate cereal crops. When adaptations at the farm level were tested (*e.g.*. change in planting date, switch of crop variety, changes in fertilizer application and irrigation), compensation for the detrimental effects of climate change was more successful in developed countries.

When the economic implications of these changes in crop yields are explored in a world food trade model, the relative ability of the world food system to absorb impacts decreases with the magnitude of the impact. Regional differences in effects remain noticeable: developed countries are expected to be less affected by climate change than developing economies. Dynamic economic adjustments can compensate for lower-impact scenarios such as the GISS and GFDL climate scenarios but not higher-impact ones such as the UKMO scenario. Prices of agricultural products are related to the magnitude of the climate change impact, and incidence of food poverty increases even in the least negative climate change scenario tested.

When the effects of lower future population and economic growth rates and trade liberalization were tested in the food trade model, reduced population growth rates would have the largest effect on minimizing the impact of climate change. Lower economic growth results in tighter food supplies, and consequently would result in higher rates of food poverty.

Full trade liberalization in agriculture, on the other hand, provides for more efficient resource use and would reduce the number of people at risk from hunger by about 100 million (from the reference case of c. 640 million in 2060). However, all of the scenarios of future climate adopted in this study increase the estimates of the number of people at risk from hunger.

It should be emphasized that the results reported here are not a forecast of the future. There are very large uncertainties that preclude this: particularly the lack of information on possible climate change at the regional level, the effects of technological change on agricultural productivity, trends in demand (including population growth), and the wide array of possible adaptations. The adoption of efficient adaptation techniques is far from certain. In developing countries there may be social or technical constraints, and adaptive measures may not necessarily result in sustainable production over long timeframes. The availability of water supplies for irrigation and the costs of adaptation are both critical needs for further research.

Future trace gas emission rates, as well as when the full magnitude of their effects will be realized, are not certain, and only a limited range of GCM climate change scenarios, representing the upper end of the projected warming, was tested. However, it can be argued that the use of scenarios from the higher GCM projections provides perspective on the downside risk of global warming projections. Because of these uncertainties, the study should be considered as an exploratory assessment of the sensitivity of the world food system to a limited number of what is, in effect, a much wider array of possible futures.

Determining how countries, particularly developing countries can and will respond to reduced yields and increased costs of food is a critical research need arising from this study. Will such countries he able to import large amounts of food? From a political and social standpoint, these results show a decrease in food security in developing countries.

The study suggests that the worst situation arises from a scenario of severe climate change, low economic growth, and little farm-level adaptation. In order to minimize possible adverse consequences-production losses, food price increases, and people at risk of hunger-the way forward is to encourage the agricultural sector to continue to develop crop breeding and management programmes for heat and drought conditions (these will be immediately useful in improving productivity in marginal environments today), in combination with measures taken to slow the growth of the human population of the world. The latter step would also be consistent with efforts to slow emissions of greenhouse gases. The source of the problem, and thus the rate and eventual magnitude of global climate change.

DEVELOPMENTS IN FOOD TECHNOLOGY

Several companies in the food industry have played a role in the development of food technology. These developments have contributed greatly to the food supply. Some of these developments are:

- *Instantized Milk Powder*: D.D. Peebles (U.S. patent 2,835,586) developed the first instant milk powder, which has become the basis for a variety of new products that are rehydratable in cold water or milk. This process increases the surface area of the powdered product by partially rehydrating spray-dried milk powder.
- *Freeze Drying*: The first application of freeze drying was most likely in the pharmaceutical industry; however, a successful large-scale industrial application of the process was the development of continuous freeze drying of coffee.
- *High-Temperature Short Time Processing*: These processes for the most part are characterized by rapid heating and cooling, holding for a short time at a relatively high temperature and filling aseptically into sterile containers.
- *Decaffeination of Coffee and Tea*: Decaffeinated coffee and tea was first developed on a commercial basis in Europe around 1900. The process is described in U.S. patent 897,763. Green coffee beans are treated with steam or water to around 20 per cent moisture. The added water and heat separate the caffeine from the bean to its surface. Solvents are then used to remove the caffeine from the beans. In the 1980s, new non-organic solvent techniques have been developed for the decaffeination of coffee and tea. Carbon dioxide under supercritical conditions is one of these new techniques. U.S. patent 4,820,537 was issued to General Foods Corp. for a CO_2 decaffeination process.

Powdered milk

Powdered milk is a powder made from dried milk solids. Powdered milk has a far longer shelf life than liquid milk and does not need to be refrigerated due to its low moisture content.

History

Powdered milk was first made in the 20th century and is used extensively today. It is found abundantly in many countries because of reduced transport and storage costs (as it does not require refrigeration).

Like other dry foods it is considered non-perishable and is favoured by survivalists, hikers and other people in need of non-perishable easy to prepare foodstuffs.

Processing

Powdered milk is typically made by spray drying non-fat skim milk. Pasteurized milk is first concentrated in an evaporator to about 50 per cent milk solids. The resulting concentrated milk is sprayed into a heated chamber where the water almost instantly evaporates, leaving behind fine particles of powdered milk solids. Alternatively, the milk can be dried by drum drying. Milk is applied as a thin film to the surface of a heated drum, and the dried milk solids are then scraped off with a knife. Powdered milk made by drum drying tends to have a cooked flavour, due to caramelization caused by greater exposure to heat.

Yet another process is the use of freeze drying, freeze drying has the advantage of preserving many nutrients in milk compared to drum drying.

Uses

Powdered milk is often used in baking, in recipes where adding liquid milk would render the product too thin to be used. It is also a common sight in UN food aid supplies, fallout shelter, warehouses and wherever fresh milk is not a viable option.

Trivia

"Powdered Milk Man" is a supervillain, mostly described in a song from The Aquabats.

Freeze drying

In a typical phase diagram, the boundary between gas and liquid runs from the triple point to the critical point. Freeze drying (blue arrow) brings the system around the triple point, avoiding the direct liquid-gas transition seen in ordinary drying (green arrow).

Freeze drying (also known as lyophilization) is a dehydration process typically used to preserve a perishable material or make the material more convenient for transport. Freeze drying works by freezing the material and then reducing the surrounding pressure to allow the frozen water in the material to sublimate directly from the solid phase to gas in a manner similar to that which causes unused ice cubes to shrink in a frost-free freezer. The greatly reduced water content that results inhibits the action of microorganisms and enzymes that would normally spoil or degrade the substance.

The application of high vacuum in freeze drying causes ice to sublimates much more quickly, making it useful as a deliberate drying process. A cold condenser chamber and/or condenser plates provide a surface(s) for the vapour to re-solidify on. These surfaces must be colder than the temperature of the surface of the material being dried, or the vapour will not migrate to the collector. Temperatures for this ice collection are typically below -50 °C.

If a freeze-dried substance is sealed to prevent the reabsorption of moisture, the substance may be stored at room temperature without refrigeration, and be protected against spoilage for many years. Freeze drying tends to damage the tissue being dehydrated less than other dehydration methods, which involve higher temperatures. Freeze drying doesn't usually cause shrinkage or toughening of the material being dried, and flavours/smells also remain virtually unchanged.

Liquid solutions that are freeze-dried can be rehydrated (reconstituted) much more quickly and easily because it leaves microscopic pores in the resulting powder. The pores are created by the ice crystals that sublimate, leaving gaps or pores in its place. This is especially important when it comes to pharmaceutical uses. Lyophilization also increases the shelf life of drugs for many years.

The process has been popularized in the form of freeze dried ice cream and as an example of astronaut food. It is also popular and convenient for hikers because the reduced weight allows them to carry more food and reconstitute it with available water. Freeze drying is used in the manufacture of instant coffee as well as some pharmaceuticals.

In high altitude environments, the low temperatures and pressures can sometimes produce natural mummies by a process of freeze-drying.

In chemical synthesis, products are often lyophilized to make them more manageable or more easy to dissolve in water for subsequent use.

Ultra-high-temperature processing

Ultra-high temperature processing (or UHT) is the partial sterilization of food by heating it for a short time, around 1-2 seconds, at a temperature exceeding 135°C (275°F), which is the temperature required to kill spores in milk. The high temperature also reduces the processing time, thereby reducing the spoiling of nutrients. The most common UHT product is milk, but the process is also used for fruit juices, cream, yoghurt, wine, soups, and stews. UHT milk has seen large success in Europe, 7 out of 10 Europeans drink it regularly. Its largest manufacturer, Parmalat, had $6 billion of sales in 1999. In the North American market, consumers have been uneasy about buying warm milk, and have been much more reluctant in buying it. To combat this, Parmalat is developing UHT milk in old-fashioned containers. Many milk products in North American foods are made using UHT milk anyway, such as McDonalds McFlurries.

UHT milk has a typical shelf life of six to nine months, until opened. However, once contaminated, UHT milk spoils much more rapidly than pasteurized milk because all the lactobacillus in milk has been killed by the high temperature heat treatment. Many people believe that UHT products are inferior in quality to less-aggressively-pasteurized alternatives, but this is disputed by the manufacturers. It can be contrasted with HTST pasteurization (high temperature/short time), in which the milk is heated to 72°C (161.6°F) for at least 15 seconds.

Decaffeination

Decaffeination is the act of removing caffeine from coffee beans and tea leaves. All decaffeination processes are performed on unroasted (green) coffee beans, but the methods vary somewhat. They generally start by steaming the beans. The beans are then rinsed in some solvent that contains as much of the chemical composition of coffee as possible without also containing the caffeine in a soluble form.

The process is repeated anywhere from 8 to 12 times until it meets either the international standard of having removed 97 per cent of the caffeine in the beans or the EU standard of having less than 0.10 per cent caffeine by mass in the coffee at the end of the process. Coffee contains over 400 chemicals important to the taste and aroma of the final drink; this effectively means that no chemical reaction will remove only caffeine while leaving the other chemicals at their original concentrations. While they are occasionally referred to informally as "decaffeinated," soft drinks without caffeine are prepared by simply leaving caffeine out in the first place.

Coffea arabica normally contains about half the caffeine of coffea robusta, and a coffea arabica bean containing a tenth as much caffeine as a normal bean has been found by some Brazilian scientists. This may change how low caffeine coffee is produced in the future. But for now, one of several methods is employed.

Roselius process

The first commercially successful decaffeination process was invented by Ludwig Roselius and Karl Wimmer in 1905. It involved steaming coffee beans with a brine (salt water) solution and then using benzene as a solvent to remove the caffeine. Coffee decaffeinated this way was sold as *Cafe sanka* in France and later as *Sanka* brand coffee in the US. Today benzene is considered so unsafe that in the 1990's the US military reformulated Napalm to reduce the benzene level.

Direct method

In the direct method the coffee beans are first steamed for 30 minutes and then repeatedly rinsed with either methylene chloride or ethyl acetate for about 10 hours. The solvent is then drained away and the beans steamed for an additional 10 hours to remove any residual solvent. Methylene chloride is considered a superior solvent because the extraction can be performed at a lower temperature and because it extracts fewer components other than the caffeine. Ethyl acetate can be obtained from various fruits and vegetables, which allows the beans processed with it to be labelled as "Natural Decafs".

Water method

In the water method (also called indirect method) beans are first soaked in hot water for several hours. Then the water is removed and either methyl chloride or ethyl acetate is used to remove the caffeine from the solution. The solution is then heated to evaporate the methyl chloride or ethyl acetate. Finally beans are soaked again in water after the caffeine is removed. The first few batches are tossed out and the water reused so that the water actually has a similar composition to the beans' liquids, except for the lack of caffeine.

CO_2 / O_2 process

With the CO_2 process, pre-steamed beans are soaked in a liquid bath of carbon dioxide at 73 to 300 atmospheres. After a thorough soaking, the pressure is reduced allowing the CO_2 to evaporate, or the pressurized CO_2 is run through either water or charcoal filters to remove the caffeine.The carbon dioxide is then used on another batch of beans. This same process can also be done with oxygen (O_2). These liquids work better than water because they are kept in supercritical state near the transition from liquid to gas so that they have the high diffusion of gas and the high density of a liquid.

Filtered water method

This is the same as the Water Method, except a charcoal filter is used instead of methyl chloride or ethyl acetate to remove the caffeine from the water. The charcoal is generally bathed in some carbohydrate, such as sucrose. This prevents the charcoal from absorbing many of the similar molecules in the water.

Triglyceride process

Green coffee beans are soaked in a hot water/coffee solution to draw the caffeine to the surface of the beans. Next, the beans are transferred to another container and immersed in coffee oils that were obtained from spent coffee grounds. After several hours of high temperatures, the triglycerides in the oils remove the caffeine - but not the flavour elements - from the beans. The beans are separated from the oils and dried. The caffeine is removed from the oils, which are reused to decaffeinate another batch of beans. This is a direct contact method of decaffeination.

THE FOOD CONTAINER INDUSTRY

Improved food technology in the United States during the late nineteenth century was brought about by two main forces. First, was the necessity for preparing for the lean years and the unproductive seasons. Second, was the need to supply the rapidly growing urban population. To meet these demands, new technological processes to prevent spoilage were devised. Methods for supplementing the ageold practice of preserving food by cold storage, by pickling in brine, or by using spices had already been introduced during the first half of the nineteenth century. The last half of the century witnessed the introduction of a number of mechanical devices used in preparing the food for processing, and machines for making cans used as containers.

Food Canning. In the earlier period, food preservation in tincoated cans had been attempted with some success, and the manufacture of cans had developed into a separate industry. During the 1850's a machine press had been invented for cutting tops and bottoms for the cans, and a die was introduced that produced a can bottom in one operation. Improved tops were made by cutting a hole in the top large enough to permit the food to be put in after the top had been soldered on. This was called the "hole-in-top can." The round disc, or "cap," was soldered over the hole after the processing had been accomplished.

Tillery Copper. Following the Civil War, several improvements in this type were made. In 1876 the Howe floater eliminated the tinner's soldering iron, since a machine rolled the cans at an angle in a bath of solder. The Tillery copper was even more efficient. It spread solder around the edge of the cap in one operation. In 1887 the Cox capper transformed the process into mass production by operating on six cans at once, completing a tray of twenty-four cans automatically.

Canmaking had become a machine production, and the industry had become largely an independent one with the canners purchasing their cans instead of making them themselves.

Automatic Machines. More improvements were added, so that by 1900 the entire operation of canmaking was one continuous line. In 1880, with the introduction of canmaking machines, two men and a couple of boys could turn out fifteen hundred cans a day. By 1900 automatic machinery was turning out cans at the rate of fifty a minute.

Sanitary Can. But canmaking had not yet reached the perfection the trade demanded. The soldering operation was objectionable, because of the popular prejudice against solder and because the canners resented destroying the attractiveness of their fancy fruits by crushing them into the small hole-in-the-top cans. The problem was solved near the close of the century when a machine was devised that mechanically crimped the top, fitted it over the flanged edge of the body, and thereby interlocked it. With a grooved roll operating at great speed, the layers of tin were pressed together. A closing machine completed the operation. The fruit could be inserted into the can without breaking, and the can was solderless. The Sanitary Can Company was organized soon after

1900 to produce this type of container. It was not until around 1920, however, that the sanitary can was generally received and the old hole-in-top type was discarded.

Machines Prepare Food for Canning. Machines for the preparation of food for canning were equally important as the machines for making cans. When it was discovered that fruits and vegetables should be processed as soon as possible after they were picked, this fact, together with the shortage of labour, led to the invention of machines to expedite the process. The local corn-husking and peashelling "bees" of former years were supplemented by itinerant labourers, not always dependable or skilled in the art. As new machinery was devised, canneries were often moved to the centres of crop production, which obviated the necessity for the long haul. But the employment of the itinerant workers, together with their housing problem, had become a great social problem. The labour-saving machinery in all phases of canning became more necessary as canneries came to be located in areas remote from centres of population.

Barkley's Machine. With canned corn always in demand—America has often been described as a corn-fed nation—the preparation of corn brought forth a number of inventions. Isaac Winslow's corn knife of the earlier period had given way to Volney Barkley's machine by 1875. This was a crude contrivance, consisting of a castiron frame about 4 feet long, attached to a trough. Four spiral knives were fitted in one end. An ear of corn was placed in the trough and forced against the knives by applying a plunger. The machine, operated by one man and a boy, could cut the kernels from 3,000 pounds of corn in 10 hours.

Warfield and Sprague Cutters. The process was further improved by the Warfield cutter, which cut a bushel of corn a minute. Streamlined production was next inaugurated with the continuous chainfeed cutter invented by Welcome Sprague, which came into use in the early 1890's. Resilient wire springs were needed to hold the ears, but because wire mills were not at that time turning out the quality of wire needed, Sprague substituted rubber teething rings for the purpose. A Sprague machine could cut sufficient corn in one day to pack fifteen thousand cans. An automatic corn husker completed the mechanization of the preparation of corn. This was fashioned somewhat like a washing machine wringer. It automatically grasped six ears at a time and ripped the husks off at the rate of three thousand ears a minute, and it was practically independent of human hands.

Pea Sheller. Canned peas, often called the "foreign aristocrat" among our vegetables, were for many years imported chiefly from France. In preparing peas for canning, hand picking and hand shelling were the chief handicaps, until the Scott and Chisholm pea sheller came into use in the 1880's. This machine removed peas from the pods as fast as six hundred hand workers could shell them. The task of removing the pods from the vines, another handicap to rapid canning, was solved during the 1890's when a combination podder and viner was invented.

Pea Podder and Viner. Meantime, horticulturists had produced a pea that matured more uniformly, and thus permitted simultaneous picking of the entire crop. Harvesting machines, loaded with racks similar to hayracks, moved into the fields, hauled the peas to a nearby factory equipped with a combined podder and viner, into which the peas were pitched like so much hay. Here they were shelled, sorted and sized, and the pods and vines were discarded for the silo. The podder and viner were major technological contributions. Within a few years they were supplanting the hand work of thousands of persons; and by 1900, some two hundred and fifty mechanized viners were performing the work that formerly would have required the services of fifty thousand labourers.

Tomatoes. Tomatoes and tomato products led the list of canned vegetables. Although Harrison Woodhull Crosby, chief gardener at Lafayette College, first successfully processed tomatoes commercially in the 1840's, the superstitious belief that tomatoes (love apples) were poisonous kept the public from accepting the product. It was not until the 1870's that tomatoes became one of the "big three" in canned vegetables. Tomato juice, however, continued to be a nuisance in the trade until the 1920's.

Fruits. Other fruits and vegetables soon became an important part of the canning industry. Peaches, which had been a product of the eastern states, were introduced into California, where the cling variety became an important article for canners. The lye peeling method, borrowed from the technology of the Indians, was so adaptable to the preparation of cling peaches that it long kept the canning of the free-stone variety out of the industry. Mechanical peach pitters and cherry pitters were introduced and became great labour-saving devices. Pineapples came into the market with the establishment of a cannery in Hawaii in 1892, but only on a small scale.

Canned Salmon "Iron Chink." One of the most remarkable mechanical aids in preparing another food product for canning was the invention of a small machine called the "iron chink." It was used in the salmon industry, which did the work of butchering the salmon and preparing them for the canneries, formerly done by Chinese on the west coast. An endless belt conveyed the salmon to the machine. The machine adjusted itself to the size and dimensions of each salmon, removed the head, tail, and fins, slit open the body, discarded the entrails, and conveyed the edible portions to the canner. It could process seventy to eighty salmon a minute.

Assembly Line Production. More important, however, in the history of food technology than any of these labour-saving devices, was the organization of the continuous-line system, whereby the canning process was carried on in one automatic mechanical operation. The factory operations in canning include preparing the product for canning, washing, peeling, grading, filling, cooking, cooling, packaging, and warehousing. Each process was a separate time-consuming task until a method was developed for streamlining all the opera. tions. Daniel Trench of Chicago combined a corn cutter, a corn filler, a rotary capping machine, pressure kettles, hoists, and conveyors in one continuous production line.

Automatic Canning. As early as 1883 the Norton brothers began the establishment of a complete automatic line for the manufacture of containers. A factory in Indiana, a canning-center state, perfected an endless chain in a processing method whereby the cans were filled and passed on a belt through a temperature of 250 degrees, then to a water cooler, and then to the warehouse without any stoppage in the process. During the operation, cans were not touched by hand.

Mass Production Lowered Cost. A number of other inventions aided in improving and refining the continuous-line process, which brought mass production in the nation's food industry, with resulting lower costs. Senator George Vest (Missouri), in noting the advances in the canning of foods, declared that, "Years ago canned foods were delicacies, but today (1890) they are the poor man's food, being cheaper in the cans than in the green articles." He noted that the canning industry had become one of the largest industries in the country, and that our canned fruits, meats, and vegetables were supplying the French, German, and English armies.

7

Techniques of Food Packaging

INTRODUCTION TO FOOD PACKAGING

Food packaging is packaging for food. It requires protection, tampering resistance, and special physical, chemical, or biological needs. It also shows the product that is labelled to show any nutrition information on the food being consumed.

FOOD WRAPPING, PAPER, CARTONS, AND PLASTIC

Paper and cardboard cartons emerged at the end of the nineteenth century as material for wrapping and packaging food. For a long time the price of materials for food packaging—tinplate, glass, and, to a lesser extent, paper—remained high and was often more costly than the food itself.

Technological innovations made it possible to produce packaging material cheaply. Paper became important for wrapping food when it could be produced from wood pulp, but paper and cardboard cartons were not yet suitable for packaging fluids. In the 1880s in the United States, paper and cartons could be made impermeable to fat and fluid by coating them with a thin film of paraffin.

In the 1930s, cellophane became an important material for food packaging, but it was gradually replaced by the expanding possibilities of polyethylene and other forms of plastic.

Another breakthrough was the invention of the Tetra Pak in Sweden in 1952, which increased the capabilities of carton containers for packaging milk, fruit drinks, and other liquids.

The carton container coated with polyethylene became a serious threat to the market for glass and cans. In the 1940s, food packaging entered the era of fully disposable packaging. The convenience of the microwave was further enhanced in the 1980s with the development of special packaging materials. The demand for ready-to-eat fresh vegetables and fruits stimulated the development of Modified Atmosphere Packaging (MAP).

LABELS AND BRAND NAMES

Closely associated with the history of food packaging is the development of food labels and brands. In the first half of the nineteenth century, food manufacturers realised that their products would sell better if a brand name was attached to them, a name with prestige that potential customers could easily recognise. Initially, labels with information about the contents were put on glass containers or cans.

Gradually, the label and the packaging as a whole became a means for promoting the food product. In most industrialised countries, legislation regulates the information that must be provided on packaging for consumers' protection.

Since the beginning of the nineteenth century, food packaging has been closely associated with industrialization and urbanization. Originally, food packaging in glass and cans was primarily meant to preserve food, but convenience became the most significant aspect of food packaging in the twentieth century.

The retail revolution, when supermarket chains supplanted family-owned grocery stores, made food packaging an indispensable part of urban food culture. On the other hand, it created problems of waste disposal, a much-discussed concern of critical consumers.

TRENDS IN FOOD PACKAGING

Numerous reports of industry associations agree that use of smart indicators will increase. There are a number of different indicators with different benefits for food producers, consumers and retailers.

TEMPERATURE RECORDERS

Temperature recorders are used to monitor products shipped in a cold chain and to help validate the cold chain. Digital temperature data loggers measure and record the temperature history of food shipments.

They sometimes have temperatures displayed on the indicator or have other output (lights, etc.): The data from a shipment can be downloaded (cable, RFID, etc.) to a computer for further analysis.

These help identify if there has been temperature abuse of products and can help determine the remaining shelf life. They can also help determine the time of temperature extremes during shipment so corrective measures can be taken.

Time-Temperature Indicators

Time-Temperature Indicators integrate the time and temperature experienced by the indicator and adjacent foods. Some use chemical reactions that result in a colour change while others use the migration of a dye through a filter media. To the degree that these physical changes in the indicator match the degradation rate of the food, the indicator can help indicate probable food degradation.

RFID

Radio Frequency Identification is applied to food packages for supply chain control and have shown a significant benefit in allowing food producers and retailers create full real time visibility of their supply chain.

FUNCTIONS OF FOOD PACKAGING

Packaging has several objectives:

- *Physical Protection*: The food enclosed in the package may require protection from, among other things, shock, vibration, compression, temperature, etc.
- *Barrier Protection*: A barrier from oxygen, water vapour, dust, etc., is often required. Permeation is a critical factor in design. Some packages contain desiccants or oxygen absorbers to help extend shelf life. Modified atmospheres or controlled atmospheres are also maintained in some food packages. Keeping the contents clean, fresh, and safe for the intended shelf life is a primary function.
- *Containment or Agglomeration*: Small items are typically grouped together in one package for reasons of efficiency. Powders, and granular materials need containment.
- *Information Transmission*: Packages and labels communicate how to use, transport, recycle, or dispose of the package or product. Some types of information are required by governments.
- *Marketing*: The packaging and labels can be used by marketers to encourage potential buyers to purchase the product. Package design has been an important and constantly evolving phenomenon for several decades. Marketing communications and graphic design are applied to the surface of the package and (in many cases) the point of sale display.
- *Security*: Packaging can play an important role in reducing the security risks of shipment. Packages can be made with improved tamper resistance to deter tampering and also can have tamper-evident features to help indicate tampering. Packages can be engineered to help reduce the risks of package pilferage: Some package constructions are more resistant to pilferage and some have pilfer indicating seals. Packages may include authentication seals to help indicate that the package and contents are not counterfeit. Packages also can include anti-theft devices, such as dye-packs, RFID tags, or electronic article surveillance tags, that can be activated or detected

by devices at exit points and require specialised tools to deactivate. Using packaging in this way is a means of retail loss prevention.

- *Convenience*: Packages can have features which add convenience in distribution, handling, stacking, display, sale, opening, reclosing, use, and reuse.
- *Portion Control*: Single serving packaging has a precise amount of contents to control usage. Bulk commodities (such as salt) can be divided into packages that are a more suitable size for individual households. It also aids the control of inventory: selling sealed one-litre-bottles of milk, rather than having people bring their own bottles to fill themselves.

PACKAGING MACHINES

A choice of packaging machinery includes technical capabilities, labour requirements, worker safety, maintaina-bility, serviceability, reliability, ability to integrate into the packaging line, capital cost, floorspace, flexibility (change-over, materials, etc.), energy usage, quality of outgoing packages, qualifications (for food, phamaceuticals, etc.), throughput, efficiency, productivity, ergonomics, etc.

Packaging machines may be of the following general types:

- Blister, Skin and Vacuum Packaging Machines.
- Capping, Over-Capping, Lidding, Closing, Seaming and Sealing Machines.
- Cartoning Machines.
- Case and Tray Forming, Packing, Unpacking, Closing and Sealing Machines.
- Check Weighing Machines.
- Cleaning, Sterilizing, Cooling and Drying Machines.
- Conveying, Accumulating and Related Machines.
- Feeding, Orienting, Placing and Related Machines.
- *Filling Machines*: Handling liquid and powdered products.
- Package Filling and Closing Machines.
- Form, Fill and Seal Machines.
- Inspecting, Detecting and Checkweighing Machines.
- Palletizing, Depalletizing, Pallet Unitizing and Related Machines.
- *Product Identification*: Labelling, marking, etc.
- Wrapping Machines.
- Converting Machines.
- Other speciality machinery.

REDUCING FOOD PACKAGING

Reduced packaging and sustainable packaging are becoming more frequent. The motivations can be government regulations, consumer pressure, retailer pressure, and cost control. (Reduced packaging often saves packaging costs.) In the UK, a Local Government Association survey produced by the British Market Research Bureau, compared a range of outlets to buy 29 common food items, found that small local retailers and market traders "produced less packaging and more that could be recycled than the larger supermarkets".

DEVELOPMENT OF PACKAGING AND CANNING

In preindustrial society, packaging of food was far from being unknown. It was used for food storage at home and for transport from the production place, the farm, or workshop to the local or regional market. Examples are the transport of cereals or flour in bags, tea in wooden boxes or tinplate canisters, and oils in jars.

At the household level, people salted meat and pickled vegetables and preserved them in jars. In groceries at the end of the nineteenth century, most commodities were still unpacked and sold in bulk. Products such as tea, coffee, sugar, flour, or dried fruits were weighed out in front of the customer and wrapped in paper or put into a bag. In major cities in the 1880s, the milkman came around with a dipper and can to deliver milk, which was often dirty.

ORIGIN OF MODERN PACKAGING AND CANNING

Early methods of sealing jars included waxed paper, leather, or skin, followed by cork stoppers and wax sealers. The beginning of modern food technology started with the experiments of the French confectioner Nicolas Appert. In 1795, the French Government offered a prize of 12,000 francs to anyone who could find a way to preserve food because Napoléon Bonaparte needed to provide the military with a safe food supply.

(The requirements of providing adequate food supplies for armies and navies have been of great significance in the history of modern packaging and food preservation.) After fourteen years of experimentation, Appert developed a method for preserving foods by heating. The food, meat, or vegetables, was first cooked in open kettles and placed in glass jars. After removing as much air as possible, the jars were carefully sealed with corks wired in place and then submersed in boiling water.

Appert chose glass for the container because he believed that it was air that caused the spoilage—glass is a material least penetrated by air. It is of importance to note that, in Appert's time, it was not known that microorganisms caused food to spoil. The processes involved in food spoilage were not understood until the second half of the nineteenth century as a result of the work of scientists such as Louis Pasteur on microorganisms. In 1810 Appert published his prize-winning essay on food preservation and the French emperor Napoléon awarded the 12,000-franc prize to him. Within a year, an English version appeared in London, and the new method of preserving food in glass spread quickly to other countries.

CANNING

Two individuals in England are given credit for applying and improving Appert's invention, Bryan Donkin and Peter Durand. Bryan Donkin, an associate of John Hall's at his Dartford Iron Works, realised in 1811 that iron containers could be used instead of the fragile glass, and in 1812 the factory began to produce canned food such as meat. In 1810, Peter Durand patented the use of metal containers, which were easier to make and harder to break than glass jars. (The glass jars used by Appert frequently broke.) He covered iron cans, which were prone to rust, with a thin plating of tin (which is not adversely affected by water), and invented the "tin can". By 1813, Durand was selling canned meat to the Royal Navy. The British admiralty bought these foods as part of the medical stores for distribution to sick men as well as to supply expeditions. By 1819 canning had arrived in the United States, but no one wanted canned food until the Civil War started. In 1821, the William Underwood Company in Boston introduced commercial canning in the United States. For a long time, people regarded canned foods with suspicion, and for good reasons.

In the middle of the nineteenth century, the foods produced by the canning industry were as likely to spoil as not because of inadequate heating techniques). Then, beginning in 1868, first in the United States and later in Europe, handmade cans were replaced by machine-cut types. The new technology made it possible for giant meat-canning firms like P.D. Armour to emerge in Chicago and Cincinnati. The product, however, was packed in big, thick, clumsy red cans and was not very appetizing.

The American Gail Borden was a pioneer in food canning. In 1856 he successfully produced sweetened condensed milk in cans and was granted a patent on the process. With financial support, the New York Condensed Milk Company was established in 1857. The demand for condensed milk was at first limited, but during the American Civil War it was introduced on a large scale. The Civil War contributed significantly to the popularization of canned foods in general. The army had to be fed and the government contracted with firms to supply food.

Under difficult circumstances, people learned that canned foods such as condensed milk can be tasty and nourishing. The invention of practical can openers at the end of the nineteenth century made cans easier to open, making them even more convenient for consumers. For many years, however, the flavour of most canned food left much to be desired. On the other hand, it should be realised that products such as canned peas and salmon were usually sold to people living on the American prairies or in the urban slums in Great Britain, most of whom had never eaten the fresh product.

In addition, losses due to spoilage caused by microorganisms remained high. It was not until the end of the nineteenth century that research carried out at Massachusetts Institute of Technology made a substantial contribution to improving the keeping quality, nutritional value, and taste of food products preserved in cans and glass. In the early twentieth century, the heavy cans were replaced by those made of lighter materials, and manufacturers could stress that their products were hygienically processed and, therefore, safer to eat than the traditionally unpackaged products that had been sold in bulk.

As food technology advanced, numerous chemical additives were developed to control or speed up food processing and to increase the keeping quality of canned foods. Originally, the nutritional value of food preserved by canning was not high, mainly due to the length of time required by the heating techniques. From the 1920s onward, however, the nutritional value of canned foods gradually approached that of the fresh product, thanks to modern food technology. Finally, in the 1960s, Reynolds and Alcoa companies succeeded in making all-aluminium cans out of one piece of metal, thereby solving the problem of the weight of the cans; only the lid needed to be attached. At the same time, the invention of the rip-off closure and the pop-top lid on aluminium cans made them even more convenient, and made can openers unnecessary. For consumers, the choice between fresh or canned food became largely a question of taste, convenience and preference.

GLASS

Despite its fragility and high production costs, glass had an advantage over cans: glass is chemically inert. In a metal can, iron, tin, and even lead may interact with the water of the preserved food due to chemical or galvanic reactions (although that problem had been solved when iron was replaced by lighter material). The problem of lead contamination had been removed in 1904 when the production system of the Sanitary Can Company in New York made soldering of the can unnecessary.

Glass became a relatively cheap and convenient form of packaging in1903 when Michael J. Owen in Britain invented a semiautomatic machine for producing both jars and bottles. In the nineteenth century a major problem with glass containers had been finding a way to close a relatively expensive container without making the bottle or jar useless after it had been opened. Glass bottles could be closed with a cork, but closing bottles and jars that had wide mouths remained a problem. Numerous ingenious inventions and innovations sought convenient ways to open and close glass containers (and cans as well). The breakthrough came with the invention of the zinc cap for the shoulder-seal jar. The most significant inventions were the Mason Jar in 1858 (named for its inventor, John Landis Mason), a glass jar with a thread at the neck that could be closed by screwing on a metal cap, and the Crown Cap for bottles, invented by William Painter in 1898. In rural households in Europe from the 1890s until about the 1950s, food preservation in jars of glass and bottles by means of Appert techniques was common, and small portable canning machines made it possible to use the new food preservation techniques in the 1930s and 1940s. As the technology of food preservation improved, however, homemade food preservation by means of salting and pickling in pots and jars of glass gradually decreased. With the invention of the home freezer, it largely disappeared.

PACKAGING TECHNOLOGY FOR PROCESSED FOODS

Modern food packaging technology brings traditional foods into a global arena which increasingly emphasizes their commercial and economic aspects. This means that food and packaging technologists become involved in the entire food supply system. This system ranges from the sea, village farm, plantation, to the markets and consumers in towns and cities, not only in their own country, but also in distant overseas markets.

The surplus foods grown in the village have a need to be more carefully harvested, protected from spoilage and damage, packaged, and transported by various means to these markets. Unless the goods are sold with minimum spoilage and at their peak flavour, appearance, and nutritional value and presented in an attractive way, they may not be eaten at all. This is a worse situation than if the crop had never been grown and can represent serious loss and waste to a community. In addition, careful emvironmental conside-rations need to be given to minimum packaging forms to avoid pollution problems and ensure sustainabilty. Very little investment has been made so far in developing traditional technologies or in applying scientific knowledge in most of the developing countries; meanwhile the more expensive products of imported technologies have further slowed the development of indigenous technologies.

It has been increasingly recognized that the time has come when these traditional technologies must be upgraded through scientific application of packaging principles and then integrated with other packaging functions such as marketing and advertisement. Technologies are called traditional if, unaffected by modernization, they have been commonly applied over a long period of time. In general, traditional technologies tend to be cheap, easy to produce, apply, maintain, and repair. They are generally labour-intensive, which can be economically beneficial, but as far as food packaging technologies are concerned, the final products are often hygienically sub-standard and they usually have a short shelf-life.

Many traditional foods have nonetheless remained unchanged in process or package for centuries, due to the fact that they developed in a particular location and are deep rooted in the natural, cultural, religious, and socio-economic environment. Some have disappeared without a trace as a result of modern influences, while others have expanded on a global scale, becoming household products, *e.g.*, soysauce, now a multimillion dollar industry. The reasons for this phenomenon need to be examined.

TRADITIONAL FOOD PACKAGING TECHNOLOGIES

Food Systems

Population drift from rural to urban areas has caused drastic changes in the food supply network from farm to the consumer in many emerging nations. One traditional belief which can no longer be sustained is the old saying that, "There are always fish in the rivers and lagoons, and rice, taro, or sweet potato in the fields; therefore let there be no concern for the next meal."

It is more likely that the farmer has gone to the city or even overseas, and is earning a labourer's wage to keep his family in food during the off-season. So food is now brought to the market by many and various means, and redistributed to these new consumers. There are new vistas for traditional food markets where the technologies are tested beyond their limits.

Given the circumstances in which many developing countries are today, the challenge for their traditional technologies is that often they do not contribute sufficiently to meeting socio-economic imperatives. This is true also of those food technologies where many of the processing methods have remained unchanged for centuries, and are becoming inadequate to cope with modem needs, because they are too labour-intensive and depend now too much on natural environmental conditions.

It is now clear that there is a need to lessen the dependence on nature, reduce the drudgery, shorten the time of the work involved and upgrade the preparation, quality, packaging, presentation, and shelf-life of these traditional foods and their packaging.

Women and Food Processing

Women play a major role in most traditional food processing and packaging. They dry leaves, pulses, and cereals, make curds and cheese, smoke meat and fish, ferment, grate, dry sweet potato and carry out a wide range of food preservation and packaging processes. It is indispensable, therefore, that women view the proposed technology improve-ments as capable of reducing their labour, without diminishing their role and status, or, in the case of marketable products, their profit. It should be borne in mind that the upgrading of traditional food and food packaging technologies is a sensitive area, for which reason the subject should be approached with caution and due regard paid to the social, economic and cultural factors involved, in addition to the gender issues.

Further sensitivity needs to be applied to the small business operation in the preparation and packaging of traditional foods. Programmes for mass production of a particular food, many have dire consequences for the small business operator. In the consideration of the structural characteristics of traditional food industries, in particular the application of new food technologies and the use of labour saving continuous large-scale processing, the task needs to be approached thoughtfully.

Upgrading of Food Packaging

Upgrading of traditional food packaging technologies in many cases, introduces exogenous factors, *i.e.*, the importation of technology from abroad. Whether or not adapted to local circumstances, the use of imported packaging technologies in many developing countries remains restricted to modern technologies; even when these are locally developed, they are more complex to use, repair, and maintain. They are also expensive and tend to rely on imported components and non-renewable sources of energy.

VACUUM PACKAGE FOOD AT HOME

There are numerous types of equipment being marketed for vacuum packaging food at home. They vary greatly in technological sophistication and price, and usually are called vacuum packaging machines or vacuum sealers. These machines may extend the storage time of refrigerated foods, dried foods and frozen foods. However, vacuum packaging is not a substitution for the heat processing of home canned foods. Vacuum packaging is also not a substitution for the refrigerator or freezer storage of foods that would otherwise require it. In fact, vacuum packaging can add to the concerns associated with storing of these perishable foods (which are foods not stable at room temperature and requiring cold storage).

There are many precautions that must be taken when vacuum packaging perishable foods for refrigerator or freezer storage. You must assume that the perishable food carries the risk of potential pathogenic contamination. And, when frozen food is ready to be thawed and used, steps to minimize the risks from microorganisms in food must still be followed. Again, perishable foods must still be refrigerated or frozen for storage after packing in a vacuum or partial vacuum environment.

Producing a vacuum means removing air from the contents of a package. Oxygen in environmental air does promote certain reactions in foods which cause deterioration of quality. For example, oxidative rancidity of fats in food and certain colour changes are promoted by the presence of oxygen. Therefore, removal of oxygen from the environment will preserve certain quality characteristics and extend the food's shelf life based on quality. However, removal of oxygen from the surrounding environment does not eliminate the possibility for all bacterial growth; it just changes the nature of what is likely to occur. In fact, what is most likely to be eliminated is growth of spoilage bacteria. The bacteria that normally spoil the quality of food in noticeable ways (odour, colour, sliminess, etc.) like to have oxygen in the environment. If able to multiply on foods, these spoilage bacteria can let you know if a food is going bad before it reaches the point it makes someone sick. In an almost oxygen-free environment like vacuum packaging produces, the spoilage bacteria do not multiply very fast so the loss of food quality is slowed down.

Some pathogenic (illness-causing) bacteria, however, like low-oxygen environments and reproduce well in vacuum-packaged foods. In fact, without competition from spoilage bacteria, some pathogens reproduce even more rapidly than in their presence. These bacteria often do not produce noticeable changes in the food, either. In the vacuum-packaged environment, food may become unsafe from pathogenic bacterial growth with no indicators to warn the consumer; the bacteria that would also normally be multiplying and spoil food in ways to make it unappealing (odour, sliminess, etc.) are not able to function without enough oxygen.

For example, *C. botulinum* (a very dangerous pathogen that causes the deadly botulism poisoning under certain conditions) grows at room temperature in low-acid moist foods if the package presents anaerobic (lacking in oxygen) conditions—if the bacteria are present, of course. Without the competition from spoilage bacteria, reproduction is even easier. Refrigeration at 38-40 °F becomes a critical step for storage of low-acid vacuum-packaged foods that aren't otherwise stable (don't keep) at room temperature (*e.g.*, canned properly). The actual temperature of the refrigerator and the temperature at which it keeps the food are essential to maintain safety of this product. If the food were not packaged under vacuum, the oxygen in the environment would offer some protection against *C. botulinum* growth and toxin development in the package.

The removal or reduction of oxygen in the storage environment is indeed helpful for extending the storage quality of non-perishable dry foods such as dried nuts or crackers. Products like this are low enough in moisture that bacterial growth is prevented. Vacuum packaging can also be safe for food that will be stored frozen. However, proper thawing under conditions that minimize bacterial growth—like refrigeration—would be essential. If the package stays closed during thawing, you still have a vacuum environment where pathogenic bacteria can be active if the temperature is warm enough.

There is no advantage to combining the use of a vacuum packaging machine with boiling water or pressure canning of foods. Jars processed in either canner develop sufficient vacuums for safe, stable storage at room temperature. They also have the added advantage of a heat process that kills pathogenic bacteria able to grow in that food at room temperature.

So is a vacuum packaging machine needed or are there advantages to owning one? One would need to ask if the amount of the investment is worth the uses for the appliance. Traditionally recommended freezing procedures and packaging methods, if carried out carefully, will produce high quality products with reasonably lengthy storage times. Storing crackers, nuts and other dried foods in air-tight storage containers will also keep them of high quality for a reasonable period of time for normal usage.

And, perishable foods still need to be treated carefully to prevent pathogens from making them unsafe. Remember, removing oxygen from a food's environment does not just solve some food storage problems—it could cause others. Consider how carefully safe food handling practices will be followed at all times, since vacuum packaging creates very good conditions for some pathogens to be a problem if any mistakes are made. For example, perishable food being vacuum packaged should not be out of refrigeration very long – no longer than 2 hours total time above 40 °F. Food that needs to be refrigerated without vacuum packaging still needs to be refrigerated! While food is being packaged and prepared or used later, extremely clean hands, and clean and sanitized equipment and work surfaces are essential. Food should be dated and still used within reasonable storage times unless frozen. Raw meats, poultry and seafood should be cooked thoroughly to recommended temperatures, measured with a food thermometer, before eating. Any food showing signs of spoilage should be discarded—*when in doubt, throw it out!*

PACKAGING RESEARCH

According to the World Packaging Organization (WPO), globally the packaging material and machinery industry is estimated to be worth $US 500 billion per year, representing between 1-2per cent of the GDP in industrialised countries. An estimated 100,000 packaging manufacturing companies employ in excess of 5 million people and, in principle, serve all business sectors manufacturing and/or trading physical products. Packaging technology has gone through a fast and significant development in recent decades; however, the smartest developments are yet to be seen.

Today's modern society depends to a large extend on the availability and use of modern packaging technology, comprising a vast variety of modern materials, high tech applications and smart operations. Modern packaging technology aims to meet a vast range of requirements ranging from providing food safety, via low cost storage and distribution, self-selling marketing, convenient consumer use to responsible waste management practices. Thanks to modern packaging technology products can be economically distributed over large geographical areas or stored over time without unacceptable quality loss. Packaging is more than a clever way of combining materials. Adding value through the development of partnership relations in the supply chain seems to be the credo for the next decennium. Good packaging facilitates a subtle cooperation between product, packaging process and material with the objective of fulfilling needs of all stakeholders along the supply chain including the post-consumer waste manager. Value chain management, product stewardship and life cycle management are considered key attributes that will drive the development of future, sustainable packaging systems. Such systems will need to go far beyond the current waste minimisation driven principles of reduce, re-use, recycle, and recovery. Packaging systems that will minimize impact on the environment, will seamlessly meet social requirements and expectations, and will be economically effective are the business winners of the future.

Drivers for Development

Today's companies in the packaging supply chain are faced with acknowledging, understanding, addressing and managing a range of issues affecting the sustainable use of packaging. Issues include the use of renewable and non-renewable resources, recyclability, regulations, and material and transport costs.

Ongoing demographic and life-style changes, technology changes, environmental issues (in particular as recognized by legislation and/or voluntary agreements in numerous countries), consumer dynamics, and supply chain demands are important factors of influence for the packaging supply chain. Adequately responding to such issues requires pro-activity, progressivism and agility from packaging manufacturers as well as users.

Despite the industry becoming more progressive and pro-active in its approach and seeking shared responsibility, the majority of the actions are still undertaken on the basis of the traditional 4-R waste management hierarchy (Reduce, Re-use, Recycle, Recover, Dispose) rather then evaluating and assessing the life cycle impacts and developing strategies to reduce impact. As a result "down gauging" and "recyclability" are still the main drivers in avoiding environmental impact.

There is little doubt that this approach will lead to a reduction of impacts in the short term. It may take the existing "waste fat" out of current packaging systems but does not take into account the need to reduce overall environmental impacts in a substantial way. It also fails to address the increasing need for packaging systems that meet requirements from new distribution systems, increasing demands for product convenience, increasing consumer differentiation and so on.

In the longer term this approach will not be sufficient and will fail to provide adequate solutions. Numerous problem areas will emerge that will require a rigorous overhaul of the use of packaging systems in order to meet ongoing commercial demands.

Particularly for fast moving consumer goods, such as food products, the packaging is one of the key product components that can provide a commercial advantage over competing products. Hence, the packaging is of significant commercial importance for the economical sustainability and growth of businesses.

There is little value in arguing the need for less packaging while economic growth is predominantly driven by diversification of markets and subsequent product variations. Key challenges for future business growth and development are:

- The ability to meet supply chain and market requirements in terms of distribution efficiency, marketing power, consumer safety and convenience, and environmental performance.
- To maintain high levels of flexibility for creating commercial advantage through value added packaging systems.
- To maximise the triple bottom line performance in order to satisfy both commercial stakeholders (shareholders, customers) and community stakeholders (government, consumers, NGOs).

These challenges cannot be successfully tackled with the traditional 4-R approach. A holistic, integrated and collabourative initiative involving the entire supply chain is essential to be able to create step change improvements.

The focus should not be on how the supply chain can reduce the amount and increase the recycling of packaging used but on how it can sustainable satisfy the economic, social and environmental requirements for packaging related to the production, distribution and consumption of products in order to further enhance the well-being of our society.

What Makes a Packaging Research

Of course, in principle a packaging (system) is just a clever way of constructing a container out of a selected material or combination of materials. A wide variety and choice is available, and selection is not an easy job.

A range of parameters, varying from product characteri-stics to consumer (client) requirements and trends, affect this selection. These parameters can be grouped in three categories as is illustrated in fig.

- Parameters in the micro or product environment.
- Parameters in the ambient or distribution environment.
- Parameters in the macro or market environment.

The parameters in the macro environment are constantly subject to change and have, to a certain extent, an effect on the ambient environment (*e.g.*, a change in distribution method can have an impact on mechanical impacts exerted on the packaging system). Because of this dynamic environment, packaging systems are continuously due to optimization, a permanent search for the optimum between functionality and cost.

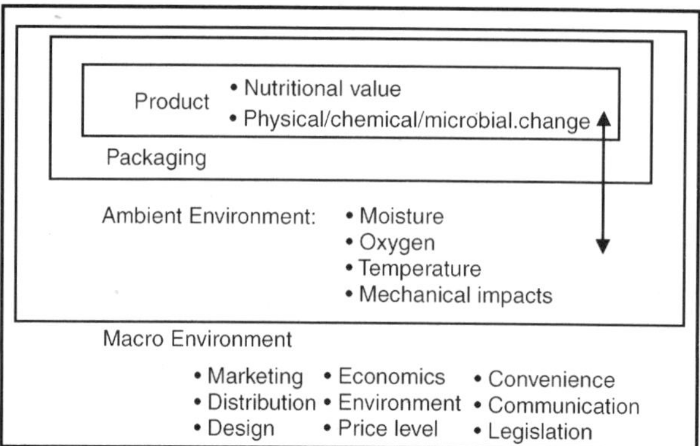

Fig. Interaction of a Packaged Food Product with the Environment.

The parameters in the macro environment are constantly subject to change and have, to a certain extent, an effect on the ambient environment (*e.g.*, a change in distribution method can have an impact on mechanical impacts exerted on the packaging system). Because of this dynamic environment, packaging systems are continuously due to optimization, a permanent search for the optimum between functionality and cost.

This value analysis includes relating the packaging system's technical, economical and environmental performance to requirements from product, manufacturing and packaging process, warehousing and distribution, retailing and marketing operations, consumer demands and behaviour, and post-consumer waste management.

Meeting this complex of varying and often opposing demands is obviously not an easy task and requires a thorough understanding of issues involved and ability to balance them in anticipation on the pull of a changing market. Packaging research and education can and will support the packaging chain in their efforts to evaluate the complexity of demands and create adequate solutions. Research in the packaging area can principally be distinguished into three main areas:

- Strategic Research
- Applied Research
- Fundamental Research.

Strategic research has a focus for new approaches in the packaging domain. It aims at identifying new pathways for advancing the future of packaging. Strategic research is essential to identify and understand future issues, to identify potential solutions (being it technological, organisational or managerial), and to design routes for implementation.

The packaging domain cannot advance without a substantial strategic research being undertaken. As in every other discipline novel approaches are essential to achieve effective future packaging solutions. However, the extensive complexity of the field due to its multi-disciplinarity and material variety, its range of stakeholders, and its broad mix of functions, makes strategic research indispensable for the packaging domain. On an individual member basis and, even more relevant, by collabouration between members, IAPRI is well positioned to undertake this type of research. By combining academic with industry focussed research strengths, IAPRI is able to provide unique sets of research skills to assist both industry and government stakeholders in the strategic advancement of packaging.

Applied research is considered the historical scene of packaging research and has often focussed on specific problems concerning packaging intensive products. The challenge ahead lies in the conversion of disciplinary and/or technological advances achieved in other areas (*e.g.* microelectronics) to the packaging field. Particularly in advancing packaging towards active, smart and intelligent systems able to interact at different points and with various stakeholders along the supply chain, significant applied research is required.

Often this type of research involves development of "proof of concept" and "route to market" and is therefore likely to be undertaken in close cooperation with commercial entities. Most IAPRI member organisations are well experienced in conducting this type of research.

However, each member will have limitations in its scope of research for which reason collabouration between two of more members might be preferable to cover a particular applied research project. Fundamental research focuses on developing the underlying sciences for packaging material applications and technologies. This area belongs historically to the discipline areas.

Fundamental research in the packaging domain historically has focussed on understanding material and packaging system behaviour. Interaction of packaging materials in contact with product (*e.g.*, food or dangerous good) is an example of such research. Mathematical modelling of the dynamic behaviour of packaging cushioning materials is another example. Fundamental research is often the domain of research students and is an important part of generating research staff for packaging research organisations. Most IAPRI members, being either University entities or having collabourative links with relevant university entities, are involved in this kind of research.

Current and Future Packaging Technologies

Since packaging is an essential part of the modern food supply chain, it is worthwhile to consider the role of packaging in food innovation and the impact modern trends in packaging science will have on the future food industry.

Most, if not all, changes in the procedures of food processing and distribution have arisen from or been accompanied by changes in packaging. An early example is thermal processing. Although attributed to Nicholas Appert in the 1790s, it was not until Peter Durand adapted the tinplated steel can to withstand thermal processing that it became the widely practiced food process we see today.

Few foods arrive in our home un-packaged. Even so-called "fresh fruits and vegetables" need to be placed into a shopping bag; often more than one package to ensure its safe trip home in the boot of the car. While packaging is critical in delivering safe food to the consumer it represents about 25per cent of the total cost of a food. This puts manufacturers under considerable pressure to reduce costs while ensuring a safe supply of food.

The food packaging industry has been efficient in its adaptation of developments in material science to design packaging to protect foods and deliver benefits to the consumer in terms of convenience, cost, communication and choice. Innovation has been the essential ingredient in the modern packaging landscape.

This article discusses the drivers that have made major changes in food packaging and speculates on developments that will deliver future packaging. Current trends in packaging provide a useful base from which to explore opportunities that packaging offers to future consumers. The trends that will be discussed are in packaging materials, package size, environmental impact and in the intrinsic role of packaging.

Packaging Materials

Most commonly used food packaging materials are derived from petroleum products. The recent increase in the price of petroleum, which is predicted to continue for some time, adversely affects packaging manufacturers. Wherever possible, the amount of packaging per unit of food needs to be reduced.

The polyolefins, polyethylene and polypropylene, dominate food packaging because of their strength, inertness and ease of production. The development of metallocene catalysed polyethylene has enabled manufacturers to reduce the gauge of the films which will in part address the increasing costs and environmental waste.

Metallocene based polymers have other benefits such as easier processing, better clarity, fewer taints as well as ease of use and greater seal strength. Metallocene catalyzed polypropylene, which has a much greater potential to improve the properties of polypropylene films, which has come onto the market much more slowly than polyethylene, offers food manufacturers greater flexibility and heat stability than current polypropylene films.

The current dominance of polyethylene terephthalate (PET) for beverage bottles could be challenged by these polymers. The poor gas barrier properties of polypropylene, which has limited its penetration into the beverage market, is the focus of intense research world wide. The use of nanocomposites to improve barrier properties without compromising clarity offers a way to increase polypropylene's penetration. Similarly active packaging technologies such as $ZerO_2$ are innovations that can potentially provide excellent barrier properties by effectively removing any oxygen migrating through the polymer.

Packaging Size

Despite the increasing cost of packaging materials the changing consumer demographic has increased the demand for individual-serve packaging. Simple maths shows that such a trend will increase the package/food ratio. Packaging size has been decreasing for many years and will continue to do so because consumers demand convenience. Having to empty part of the contents of a package before heating is not convenient.

This creates an opportunity for innovative manufacturers to supply food in packaging that eliminates this chore. The recent appearance of self-venting microwaveable packaging for shelf stable foods exemplifies how packaging innovations allow food manufacturers to capture benefit from changing demographics.

Environmental Impact

An important consideration of any technology is the environmental impact. Packaging has been such a key factor in the evolution of the food supply chain that its environmental impact should come under scrutiny.

Environmentalists and consumers are concerned about the impact of large quantities of waste packaging material on our water systems and land fills. Since no real alternative to packaging is likely it is important that the environmental impact of packaging is minimized.

This is an opportunity for food and packaging manufacturers to meet the consumer and regulatory demand. Since 1999 the Australian industry has adopted a National Packaging Covenant as the main vehicle to address the environmental impact of packaging. It is based on the principles of shared responsibility through the supply chain. In 2005 the Covenant was revised to incorporate a national recycling target of 65per cent for packaging and no further increases in packaging waste disposed to landfill by the end of 2010. These targets offer further opportunities for packaging innovation such as bio-based polymers.

Bio-based Polymers

Biodegradable polymers like polyglycolic acid, polylactic acid (PLA), and polycaprolactone do not yet have all the properties to replace common polymers. Improvements in polycaprolactone have been achieved by blending with clay based nanocomposites.

PLA based polymers include a polyester family of compostable packaging material, which potentially offers consumers good performance and cost effectiveness. One of the limitations to the commercial success of active and intelligent packaging has been their integration into common packaging materials. Polyolefins do not readily lend themselves to the chemistries that are involved in smart packaging. The trend to bio-based packaging technologies which are much more amenable will lead to future active and intelligent packaging developments.

Functional Packaging

Recent developments in the developing functional foods market have extended into the packing of the food product. Rather than incorporating the delicate and expensive functional ingredients into the food itself, developers have begun to package the functional ingredients in the packaging itself.

Examples of such innovation include drinking straws that release the ingredient as you drink through them and bottle caps that have a sealed dose of functional ingredient that is incorporated into the drink immediately prior to consumption.

Intrinsic Role of Packaging

The role of packaging is changing. The basic and essential role of packaging is to protect and preserve the safety and nutritional quality of food. However, packaging is increasingly required to inform the consumer of the properties of the food and its package. The date code tells consumers how long they have before the food deteriorates, the label informs the consumer of the amounts of energy, fat, protein, carbohydrate and sodium content as well as potential allergens. Information on the ingredients, serving instructions and country of origin is also included. The package is now evolving to be a vital point of communication with the customer and is becoming integrated into the marketing function.

Electronics

As the cost and size of electronic components continues to decrease the use of wireless devices such as RFID tags will become commonplace and the opportunity to incorporate other functions such as temperature measurement and real-time shelf life information becomes possible. Many sensory attributes could be monitored and displayed for the consumer. The packaging landscape has undergone dramatic changes over the past 50 years, and these changes are likely to continue for the next fifty. Those companies that fail to integrate these innovations into their business plans risk failure.

Nanocomposite Polymer Films Containing Carvacrol for Antimicrobial Active Packaging

Active food packaging is an innovative solution to meet the continuous changes in current consumer demands and market trends. Active food packaging extends shelf life and improves safety of the food by scavenging of oxygen, moisture or ethylene, and by promoting emission of ethanol, flavours, and antimicrobial agents.

Antimicrobial packaging is attracting increasing attention from the food and packaging industry, since the use of preservative packaging films offers several advantages compared with the direct addition of preservatives into food products. The incorporation of antimicrobial agents into polymeric films allows industry to combine the preservative functions of antimicrobials with the protective functions of the pre-existing packaging concepts.

Antimicrobial activity can be achieved by including pads containing volatile antimicrobial agents into packages, incorporating antimicrobial agents directly into polymers, coating antimicrobials onto polymer surfaces, immobilizing antimicrobials by chemical grafting, using polymers that are antimicrobial by themselves.

A wide number of generally recognized as safe (GRAS) antimicrobials, such as silver substituted zeolites, enzymes, antioxidant phenolics, are used for food packaging applications. In recent years there has been a growing demand for the use of natural additives, including natural extracts such as essential oils, which are categorized as GRAS by US Food and Drug Administration.

The antimicrobial activity of several essential oils and the efficiency of their incorporation into plastic matrices, such as polypropylene (PP) and polyethylene (PEVethylene vinyl alcohol copolymer (EVOH) have been already reported. Carvacrol, which is present in high concentrations in thyme and oregano oil, exhibits a significant in vitro antibacterial activity. This phenolic compound has also antifungal, insecticidal. antitoxigenic. and antiparasitic activities.

Besides bacterial contamination, oxidation of food constituents as well as spoilage by moulds in presence of oxygen can cause food deterioration. In order to hinder the oxygen permeation, the use of nanocomposites represents a powerful tool for packaging industry.

It is known that the addition of nanosized lamellar clays with large aspect ratio into a polymer matrix is expected to improve polymer barrier properties at relatively low filler loadings. Since the pioneering studies carried out by Okada *et al.*, an impressive number of articles dealing with the characterization of the physical properties of polymer-clay nanocomposites have been reported, some of them related to materials suitable for food contact.

Although the observed decrease of permeability is generally ascribed to the high aspect ratio of the clay layers which should increase the tortuosity factor, the influence of nanoparticles on the crystallization rate and extent of the polymer melt also plays a role. Since crystals are generally impermeable to the transport of low molecular weight substances, the combination of higher crystallinity and impermeable clay elements may result in improved barrier properties for these materials. On the other hand, the small clay loading does not affect the processability and does not substantially increase the density of the polymer. Furthermore, due to the fine particle size of the filler, the film clarity for all practical purposes is not reduced.

This work represents a preliminary study aimed to develop antimicrobial films of low-density polyethylene (LDPE) containing carvacrol. As reported in literature, one drawback of PE liners in barrier packages for, *e.g.*, aseptically processed juices, is the loss of flavour/aroma due to the scalping effect of the polymer membrane. Nanoparticles [*i.e.*, organo-modified montmorillonite (MMT)] were used to improve the barrier properties of packaging films and to protect the volatile and heat-sensitive carvacrol from degradation. In this frame, the nanoclay filler is believed to positively interact with carvacrol dispersed into clay galleries, thus preserving the antimicrobial activity of the doped film.

8

Fruit and Vegetable Processing

IMPORTANCE OF FRUIT AND VEGETABLES

In developing countries agriculture is the mainstay of the economy. As such, it should be no surprise that agricultural industries and related activities can account for a considerable proportion of their output. Of the various types of activities that can be termed as agriculturally based, fruit and vegetable processing are among the most important.

Both established and planned fruit and vegetable processing projects aim at solving a very clearly identified development problem. This is that due to insufficient demand, weak infrastructure, poor transportation and perishable nature of the crops, the grower sustains substantial losses. During the post-harvest glut, the loss is considerable and often some of the produce has to be fed to animals or allowed to rot.

Even established fruit and vegetable canning factories or small/medium scale processing centres suffer huge loss due to erratic supplies. The grower may like to sell his produce in the open market directly to the consumer, or the produce may not be of high enough quality to process even though it might be good enough for the table. This means that processing capacities will be seriously underexploited.

The main objective of fruit and vegetable processing is to supply wholesome, safe, nutritious and acceptable food to consumers throughout the year. Fruit and vegetable processing projects also aim to replace imported products like squash, yams, tomato sauces, pickles, etc., besides earning foreign exchange by exporting finished or semi-processed products.

The fruit and vegetable processing activities have been set up, or have to be established in developing countries for one or other of the following reasons:

- Diversification of the economy, in order to reduce present dependence on one export commodity;
- Government industrialisation policy;
- Reduction of imports and meeting export demands;
- Stimulate agricultural production by obtaining marketable products;

- Generate both rural and urban employment;
- Reduce fruit and vegetable losses;
- Improve farmers' nutrition by allowing them to consume their own processed fruit and vegetables during the off-season;
- Generate new sources of income for farmers/artisans;
- Develop new value-added products.

HEAT PRESERVATION/HEAT PROCESSING

Various Degrees of Preservation

There are various degrees of preservation by heating; a few terms have to be identified and understood.

- Sterilisation. By sterilisation we mean complete destruction of micro-organisms. Because of the resistance of certain bacterial spores to heat, this frequently means a treatment of at least 121° C (250° F) of wet heat for 15 minutes or its equivalent. It also means that every particle of the food must receive this heat treatment. If a can of food is to be sterilised, then immersing it into a 121° C pressure cooker or retort for the 15 minutes will not be sufficient because of relatively slow rate of heat transfer through the food in the can to the most distant point.
- "Commercially sterile". Term describes the condition that exists in most of canned or bottled products manufactured under Good Manufacturing Practices procedures and methods; these products generally have a shelf-life of two years or more.
- Pasteurized means a comparatively low order of heat treatment, generally at a temperature below the boiling point of water. The more general objective of pasteurization is to extend product shelf-life from a microbial and enzymatic point of view; this is the objective when fruit or vegetable juices and certain other foods are pasteurized.
- Pasteurization is frequently combined with another means of preservation - concentration, chemical, acidification, etc.
- Blanching is a type of pasteurization usually applied to vegetables mainly to inactivate natural food enzymes. Depending on its severity, blanching will also destroy some microorganisms.

Determining Heat Treatment/Thermal Processing Steps

Since heat sufficient to destroy micro-organisms and food enzymes also usually has adverse effects on other properties of foods, in practice the minimum possible heat treatment should be used which can guarantee freedom from pathogens and toxins and give the desired storage life; these aims will determine the choice of heat treatment. In order to safely preserve foods using heat treatment, the following must be known:

- What time-temperature combination is required to inactivate the most heat resistant pathogens and spoilage organisms in one particular food?
- What are the heat penetration characteristics in one particular food, including the can or container of choice if it is packaged?

Preservation processes must provide the heat treatment which will ensure that the remotest particle of food in a batch or within a container will reach a sufficient temperature, for a sufficient time, to inactivate both the most resistant pathogen and the most resistant spoilage organisms if it is to achieve sterility or "commercial sterility", and to inactivate the most heat resistant pathogen if pasteurization for public health purposes is the goal.

Different foods will support growth of different pathogens and different spoilage organisms so the target will vary depending upon the food to be heated.

Technological principles of pasteurization

Physical and chemical factors which influence pasteurization process are the following:

- Temperature and time;
- Acidity of the products;
- Air remaining in containers.

Pasteurization processes. In pasteurising certain acid juices for example, there are two categories of processes:

- Low pasteurization where pasteurization time is in the order of minutes and related to the temperature used; two typical temperature/time combinations are as following:
- 63° C to 65° C over 30 minutes, or
- 75° C over 8 to 10 minutes.

Pasteurization temperature and time will vary according to:

- Nature of product; initial degree of contamination;
- Pasteurized product storage conditions and shelf life required.

In this first category of pasteurization processes it is possible to define three phases:

- Heating to a fixed temperature;
- Maintaining this temperature over the established time period (= pasteurization time);
- Cooling the pasteurized products: natural (slow) or forced cooling.
- Rapid, high or flash pasteurization is characterized by a pasteurization time in the order of seconds and temperatures of about 85° to 90° C or more, depending on holding time. Typical temperature/time combinations are as follows:

- 88° C (190° F) for 1 minute;
- 100° C for 12 seconds;
- 121°C for 2 seconds.

While bacterial destruction is very nearly equivalent in low and in high pasteurization processes, the 121° C/2 seconds treatment give the best quality products in respect of flavour and vitamin retention. Such short holding times, however, require special equipment which is more difficult to design and generally is more expensive than the 63-65 ° C/30 minutes type of processing equipment.

In flash pasteurization the product is heated up rapidly to pasteurization temperature, maintained at this temperature for the required time, then rapidly cooled down to the temperature for filling, which will be performed in aseptic conditions in sterile receptacles. Taking into account the short time and rapid performance of this operation, flash pasteurization can only be achieved in continuous process, using heat exchangers.

Industrial applications of pasteurization process are mainly used as a means of preservation for fruits and vegetable juices and specially for tomato juice.

Thermopenetration

The thermopenetration problem is extremely important, especially in the case of the pasteurization of products packed in glass containers because it is the determining factor for the success of the whole operation.

During pasteurization it is necessary that a sufficient heat quantity is transferred through the receptacle walls; this is in order that the product temperature rises sufficiently to be lethal to micro-organisms throughout the product mass. The most suitable and practical method to speed up thermopenetration is the movement of receptacles during the pasteurization process.

Rapid rotation of receptacles around their axis is an efficient means to accelerate heat transfer, because this has the effect, among others of rapidly mixing the contents. The critical speed of for this movement is generally about 70 rotations per minute (RPM). This enables a more uniform heating of products, reducing heating time and organoleptic degradation. Heating may precede or follow packaging. These principles of different temperature time combinations very largely determine the design parametres for heat preservation equipment and commercial practices.

The food processor will employ no less than that heat treatment which gives the necessary degree of micro-organism destruction. This is further ensured by periodic inspection by local sanitary authorities or by the importing countries sanitary services. However, the food processor also will want to use the mildest effective heat treatment to ensure highest food quality. It is convenient to separate heat preservation practices into two broad categories: one involves heating of foods in their final containers, the other employs heat prior to packaging. The latter category includes methods that are inherently less damaging to food quality, where the food can be readily subdivided (such as liquids) for rapid heat exchange. However, these methods then require packaging under aseptic or nearly aseptic conditions to prevent or at least minimise recontamination. On the other hand, heating within the package frequently is less costly and produces quite acceptable quality with the majority of foods and most of our present canned food supply is heated in the package. In practice, therefore, most of the canned food produced locally in developing countries should be heated within the package.

BANANA AND PLANTAIN PROCESSING TECHNOLOGIES

Traditional Processing

Products: Uses and Dietary Significance

Most of the world's bananas are eaten either raw, in the ripe state, or as a cooked vegetable, and only a very small proportion are processed in order to obtain a storable product. This is true both at a traditional village level with both dessert and cooking bananas and when considering the international trade in dessert bananas.

In general, preserved products do not contribute significantly to the diet; however, in some localised areas the products are important in periods when food are scarce. Probably the most widespread and important product is flour preparation from unripe banana and plantains by sun-drying. In Uganda, dried slices known as "mutere" are prepared for storage from green bananas, the dried slices being either used directly for cooking or after grinding into a flour.

"Mutere" is used chiefly as a famine reserve and does not feature largely in the diet under normal conditions. In Gabon, plantains are sometimes made into dried slices which can be stored and used on long journeys, and plantains are used in Cameroon to prepare dried pieces which are stored and ground as needed into flour for use in cooking a paste known as "fufu". Dried green banana slices are also used in parts of South and Central America and West Indies for preparing flour.

The other nutritionally important product is beer which is a major product in Uganda, Rwanda and Burundi where green banana utilisation is particularly high.

PRESERVATION METHODS AND PROCESSES

Drying

Both ripe and unripe bananas and plantains are normally peeled and sliced before drying, although banana figs are sometimes prepared from whole ripe fruit.

Sun drying is the most widespread technique where the climate is suitable but drying in ovens or over fires is also practiced.

In west Africa, plantains are often soaked and sometimes parboiled before drying. The slices of unripe fruit are normally spread out on bamboo frameworks; or a cemented area; or on a mat; or on a swept-bare patch of earth; or on a roof; or sometimes on stones outcrops or sheets of corrugated iron.

Oven-drying of ripe bananas is practiced in Polynesia as a mean of preserving the fruits, which are then wrapped in leaves and bound tightly to store until needed. In East Africa a method has been reported that involves drying the peeled bananas on a frame placed over a fire for 24 hr before drying in the sun, to accelerate the process.

Product Stability and Storage Problems

There is little experimental data on the storage life of the traditionally made banana and plantain products.

Potential for Scaling up of Traditional Processes to Industrial Level

Many banana products are now produced on an industrial scale, including the traditional banana figs and flour, and the processing techniques are described below. One of the main problems encountered has been the susceptibility of banana products to flavour deterioration and discolouration and in the past many products reaching the market have been of poor quality.

A great deal of research has been directed to overcoming these problems, although however good the resultant products are they cannot compare in flavour and other characteristics with the fresh banana fruit. Indeed, an important constraint on the large-scale development of banana processing is the lack of demand for banana products since the fresh fruit is available throughout the year in most parts of the tropical world. The production of beer from banana and plantains has not been scaled up to an industrial level, and while an important product in localised areas of tropical Africa, the market is rapidly declining in favour of European-type brews produced locally.

Industrial Processing

Products and Uses

The main commercial products made from bananas are canned or frozen purée, dried figs, banana powder, flour, flakes, chips (crisps), canned slices and jams. Banana products can be divided roughly into two types - those for direct consumption, such as figs, and those for use in food manufacturing industry, for example purées and powder. Banana figs, or fingers as they are sometimes known, are usually whole, peeled fruit carefully dried so as to retain their shape, although sometimes the fruit is sliced or halved to facilitate drying. Banana and plantain chips (crisps) are thinly sliced pieces of fruit fried in oil and eaten as a snack like potato chips (crisps).

The main use of canned slices is in tropical fruit salads. Banana flakes are used as a flavouring or in breakfast cereals. Banana purée find use mainly in the production of baby foods. Banana flour is said to be highly digestible and is used in baby and invalid foods, but can also be used in the preparation of bread and beverages. Banana powder is used chiefly in the baking industry for the preparation and fillings for cakes and biscuits and is also used for invalid and baby foods.

Processing Technology

In general, to obtain a good-quality product from ripe-bananas the fruit is harvested green and ripened artificially under controlled conditions at the processing factory. After ripening, the banana hands are washed to remove dirt and any spray residues, and peeled. Peeling is almost always done by hand using stainless steel knives, although a mechanical peeler for ripe bananas has been developed, capable of peeling 450 Kg of fruit per hour (Banana Bulletin, 1974).

The peeling of unripe bananas and plantains is facilitated by immersing the fruit in hot water. For example, immersion in water at 70-75 ° C for 5 min. has been suggested as an aid for peeling green bananas for flour production, while the peeling of green bananas for freezing has been facilitated by immersion in water at 93° C for 30 min.

Banana Figs

Fully ripe fruits with a sugar content of about 19.5per cent are used and are treated with sulphurous acid after peeling, then dried as soon as possible afterwards. Various drying systems have been described using temperatures between 50 and 82° C for 10 to 24 hr to give a moisture content ranging from 8 to 18per cent and a yield of dried figs of 12 to 17per cent of the fresh banana on the stem.

One factory in Australia uses a solar heat collector on the roof to augment the heat used for drying bananas. Bananas can also be dried by osmotic dehydration, using a technique which involves drawing water from 1/4-in. thick banana by placing them in a sugar solution of 67 to 70 deg. Brix for 8 to 10 hr. followed by vacuum-drying at 65 to 70° C, at a vacuum of 10 mm Hg for 5 hr. The moisture content of the final products is 2.5per cent or less, much lower than that achieved by other methods.

Banana Puree

Banana purée is obtained by pulping peeled, ripe bananas and then preserving the pulp by one of three methods: canning aseptically, acidification followed by normal canning, or quick-freezing. The bulk of the world's purée is processed by the aseptic canning technique. Peeled, ripe fruits are conveyed to a pump which forces them through a plate with 1/4-in. holes, then onto a homogeniser, followed by a centrifugal de-aerator, and into a receiving tank with 29in. vacuum, where the removal of air helps prevent discolouration by oxidation.

The purée is then passed through a series of scraped surface heat exchangers where it is sterilised by steam, partially cooled, and finally brought to filling temperature. The sterilised purée is then packed aseptically into steam-sterilised cans which are closed in a steam atmosphere.

Banana Slices

Several methods for canning of banana slices in syrup are used. Best-quality slices are obtained from fruit at an early stage of ripeness. The slices are processed in a syrup of 25 deg. Brix with pH about 4.2, and in some processes calcium chloride (0.2per cent) or calcium lactate (0.5per cent) are added as firming agents.

A method for producing an intermediate-moisture banana product for sale in flexible laminate pouches has been developed. Banana slices are blanched and equilibrated in a solution containing glycerol (42.5per cent), sucrose (14.85per cent), potassium sorbate (0.45per cent), and potassium metabisulphite (0.2per cent) at 90 deg. C for 3 min. to give a moisture content of 30.2per cent.

Banana Powder

In the manufacture of banana powder, fully ripe banana pulp is converted into a paste by passing through a chopper followed by a colloid mill. A 1 or 2 per cent sodium metabisulphite solution is added to improve the colour of the final product. Spray- or drum-drying may be used, the latter being favoured as all the solids are recovered. A typical spray dryer can produce 70 kg powder per hour to give yields of 8 to 11per cent of the fresh fruit, while drum-drying gives a final yield of about 13per cent of the fresh fruit. In the latter method the moisture content is reduced to 8 to 12 per cent and then further decreased to 2 per cent by drying in a tunnel or cabinet dryer at 60° C.

Banana Flour

Production of flour has been carried out by peeling and slicing green fruit, exposure to sulphur dioxide gas, then drying in a counter-current tunnel dryer for 7 to 8 hr. with an inlet temperature of 75° C and outlet temperature of 45° C, to a moisture content of 8per cent, and finally milling.

Banana Chips (Crisps)

Typically, unripe peeled bananas are thinly sliced, immersed in a sodium or potassium metabisulphite solution, fried in hydrogenated oil at 180 to 200°

C, and dusted with salt and an antioxidant. Alternatively, slices may be dried before frying and the antioxidant and salt added with the oil. Similar processes for producing plantain chips have been developed.

Banana Beverages

In a typical process, peeled ripe fruit is cut into pieces, blanched for 2 min. in steam, pulped and pectolytic enzyme added at a concentration of 2 g enzyme per 1 kg pulp, then held at 60 to 65° C and 2.7 to 5.5 pH for 30 min.

In a simpler method, lime is used to eliminate the pectin. Calcium oxide (0.5per cent) is added to the pulp and after standing for 15 min. this is neutralised giving a yield of up to 88per cent of a clear, attractive juice. In another process banana pulp is acidified, and steam-blanched in a 28-in Hg vacuum which ensures disintegration and enzyme inactivation. The pulp is then conveyed to a screw press, the resulting purée diluted in the ratio 1:3 with water, and the pH adjusted by further addition of citric acid to 4.2 to 4.3, which yields an attractive drink when this is centrifuged and sweetened.

Jam

A small amount of jam is made commercially by boiling equal quantities of fruit and sugar together with water and lemon juice, lime juice or citric acid, until setting point is reached.

Product Stability and Spoilage Problems

All dried banana products are very hydroscopic and susceptible to flavour deterioration and discolouration, but this can be overcome to some extent by storing in moisture-proof containers and sulphiting the fruit before drying to inactivate the oxidases.

The dried products are also liable to attack by insects and moulds if not stored in dry conditions, although disinfestation after drying by heating for 1 hr to 80° C or by fumigation with methyl bromide ensures protection against attack. Banana powder is said to be stored for up to a year commercially and flakes have been stored in vacuum-sealed cans with no deterioration in moisture, colour or flavour for 12 months.

Banana chips tend to have a poor storage life and to become soft and rancid. However, chips treated with an antioxidant have been stored satisfactorily at room temperature in hermetically sealed containers up to 6 months with no development of rancidity.

Quality Control Methods

In general a good quality product is obtained if fruit is harvested at the correct stage of maturity and, where appropriate, ripened under controlled conditions. For example, in the case of banana figs, the fruit should be fully mature (sugar content of 19.5per cent or above) or the final product is liable to be tough and lacking in flavour. However, if over-ripe fruit is used, the figs tend to be sticky and dark in colour, so the fruit must be fully yellow but still firm.

For banana flour, which is prepared from unripe bananas, the fruit is harvested at three-quarters the full-ripe stage and is processed within 24 hr. prior to the onset of ripening. If less mature fruit is used, the flour tastes slightly astringent and bitter due to the tannin content. Bananas harvested between 85 and 95 days after the emergence of the inflorescence, with a pulp-to-peel ratio of about 1.7, were considered to be most suitable for the deep-fat frying.

Other criteria suggested for assessing maturity were beta-carotene and reducing sugar content, both of which increase with increasing maturity and pH which decreases as the fruit ripens, and these should be, respectively, about 2000 $\mu g/100$ g, less than 1.5per cent and 5.8 or above. Browning was found to occur if the sugar content was higher than 1.5per cent. The determination of crude fat in processed chips is also considered to be a necessary quality control measure. It is important to remove all impurities prior to processing of products, and this is done by washing to remove dirt and spray residues and control on the processing line so that substandard fruit can be removed.

Preparation Methods for Fresh Bananas and Plantains

The main ways of preparing fresh bananas for consumption are boiling or steaming, roasting or baking and frying. Boiling followed by pounding into "fufu" is also widely adopted in certain areas of the tropics.

Boiling or Steaming

Plantains and bananas are often prepared simply by boiling in water, either in their peel or after peeling, and either ripe or unripe; if unripe, the fruit is scraped thoroughly after peeling to remove all traces of fibrous material. The boiled fruit is eaten alone or more usually accompanied by a sauce. This preparation technique is widely used in West Africa.

Roasting or Baking

Unpeeled or peeled fruit, either ripe or unripe, is roasted simply by placing in the ashes of a fire or in an oven. This method is widely used in West Africa, East Africa and the South Pacific islands. For example, ripe plantains are placed unpeeled in an oven and when partly brown and tender, removed and peeled, then replaced in the oven and roasted evenly.

Frying

Ripe or unripe plantains or bananas are often peeled, sliced and cooked in oil, particularly in West Africa and in parts of South America and the West Indies.

Similar products are also made in East Africa. Typically, ripe plantains are peeled, cut into slices or split lengthways, and fried in palm oil or with groundnut oil, the pieces being served either hot with a sauce or with fried eggs, or cold as a snack.

Pounding

Pounding is a process, used particularly in West Africa, for preparing most perishable staple food crops including plantains, cassava, yams and cocoyams to obtain a paste or dough known as "fufu" (also spelled "foofoo", "foutou", "foufou"). The plantains are peeled or boiled and peeled after boiling and pounded in a wooden mortar, the resulting paste normally being eaten with soup or a spiced sauce of meat and vegetables, but sometimes after wrapping in leaves and steaming.

BANANA AND ITS MARKET

The present study is directed neither towards justification nor criticism of the United Fruit Company's operations over the more than half century of its history. We feel that it would be an essentially sterile exercise to attempt to disentangle the skein of events since 1899. Since then, a banana industry of sufficient size to assume importance in international commerce literally has been created by imaginative traders, with the United Fruit Company exercising a dominant role in the process. To reconstruct and appraise the historical record in a way that would allow recriminations to be balanced against solid accomplishments would call for an omniscient judgement that we do not feel we possess.

Rather, we have set ourselves the much more modest, though still not unambitious, task of attempting to contribute towards an understanding of what the banana industry is, of who benefits from it and by how much, and to appraise the record of the United Fruit Company in the industry as it operates today. Our perspective, then, is contemporary rather than historic. Most of our field work was completed in 1956, and our major concentration is upon operations in the year 1955, the latest year for which national accounts data for the several banana-producing countries were available in relatively complete form.

We are acutely conscious of the fact that our study deals with a field in which the ideas of most people are coloured by strongly held preconceptions based upon an interpretation of past events rather than upon an examination of the current record. It has seemed to us that a useful service could be performed by describing the present organization and procedure of banana production and marketing, by measuring what can be measured, and by limiting subjective judgements to matters not susceptible to appraisal in objective terms. Where we are forced to make value judgements, it is our hope that at least relative objectivity may be achieved through the circumstance of a joint authorship that combines North American with Latin American outlooks. More particularly, our interest in the United Fruit Company record is focused upon its impact upon the economic development process in the six Latin American republics selected for intensive study-Guatemala, Honduras, Costa Rica, Panama, Colombia, and Ecuador. These were singled out because their combined banana shipments represent close to 60 percent of the tonnage weight of world banana exports, and because about 95 percent of the bananas handled by the United Fruit Company in 1955 was produced in or purchased from these six sources. Through this sampling procedure, we could limit our field work to supportable dimension, and still cover the bulk of the world banana production for export, and a preponderant portion of United's banana procurement operations.

In concentrating upon United Fruit's contributions to economic development in these six countries, we are guided by our conviction that the widest possible diffusion of vigorous economic growth is one of the most important concerns of the contemporary free world. We are convinced that adherence to and strengthening of democratic institutions in the less developed countries of the world depends in large measure upon the demonstration that aspirations for general economic progress can be realized under free institutions. And we believe that the flow of investment capital from the capital-generating nations of high industrial development to the capital-poor countries is a major instrument for helping to stimulate balanced growth in the latter.

We start with the premise that investment capital, private or public, will not continue to flow unless it receives a return judged to be adequate. Therefore we shall examine the profitability of United Fruit investments in the six republics upon this criterion. But it seems equally clear to us that continuing hospitality for foreign private investment ventures cannot be expected to endure unless there is clear evidence that it is contributing to the development of the host nations to a degree that would not be realizable without it.

Accordingly, we shall examine the United Fruit record to see whether or not its operations present convincing evidence of satisfactory performance upon this score. Before we embark upon such an examination of the operations of the company, it is requisite that we establish a frame of reference, by presenting a picture of the world banana market as a whole.

Production and Consumption

The Food and Agriculture Organization of the United Nations estimates world banana production in 1955 at 11.6 million metric tons or about 25.7 billion pounds. It apportions about 46 percent of this to South America, 23 percent to Central America, 23 percent to Asia, 6.5 percent to Africa, and 1.5 percent to Oceania. As such, the banana crop is the fourth largest of the world's reported fruit crops, exceeded in tonnage production only by grapes (88.0 billion pounds), by citrus fruit (39.2 billion pounds), and by apples (29.1 billion pounds).

If the portion of the grape crop produced for making wine (71.4 billion pounds) rather than for consumption as fruit is deducted, and that of apples produced for cider (8.8 billion pounds), bananas displace both grapes and apples and are second only to citrus among the world's fruit crops consumed directly as food. Excluding grapes, citrus, and apples, the tonnage of bananas exceeds by a considerable margin the combined weight of the remaining important fruit crops of the world—pears, pineapples, dates, and figs (17.2 billion pounds, combined, for 1955).

Table. World Banana Production

Producing area	Million Pounds	per cent of total
Central America	5,950	23.2
South America	11,700	45.6
Asia	5,950	23.2
Africa	1,650	6.4
Oceania	400	1.6
World Total	26,650	100.0

Table given above covers a world population for 1955 estimated at 2,490 million persons. On this basis, the average consumption of bananas by each man, woman, and child was 10.3 pounds in 1955 on a stem-weight basis. Only the subcategory of citrus fruits, comprising oranges, tan gerines, and clementines shows a larger per capita consumption, 12.6 pounds; and only apples, 8.1 pounds, and grapes, 6.6 pounds, were serious rivals of the banana among the world's consumption fruits.

It is, of course, obvious that such blanket averages have little relationship to the actual pattern of fruit consumption by the world's populace. Banana production is restricted to the tropics and more particularly to humid tropic areas. For successful growth, most varieties require a temperature range between 55° and 105° F., and almost all suffer severe damage when temperatures drop below 50° to 52° F.

Since the banana cannot be grown as a seasonal crop—its cycle of development ranges from 12 to 15 months from planting to harvest—its range of cultivation is restricted to the zones in which year-round temperatures are within these extremes. It is further restricted to areas that can provide its exceptionally high moisture requirements (four to five inches monthly or 60 inches annually as a minimum) through heavy and evenly spaced rainfall or generous water supplies for irrigation.

The banana thrives best in alluvial, well-drained soils, though it will tolerate clay soils of friable consistency. Since its stalk is merely a tightly rolled cylinder of leaf sheaves, it is particularly vulnerable to blowdowns or uprooting by floods when heavy with fruit. Even relatively mild winds can seriously injure the fruit by shredding the protecting leaf structure upon which healthy fruit production depends. The areas that can grow bananas successfully upon a commercial basis must therefore provide the necessary environmental factors of temperature, moisture, soil characteristics, and freedom from damaging wind and flood recurrence.

In general, the areas that can provide a hospitable environment for banana culture are found in low-lying, high-precipitation lands between 20° North and 24° South latitude, with the extremes set by 24° North and 30° South latitude. The range of actual banana production is indicated by the areas listed in Table given above.

It might be assumed that the areas of important production would be the areas of major consumption. Yet, as indicated in Table given below, some of the broad areas of banana production are heavy banana consumers as well, while others are not. On a per capita basis, the highest level of consumption in 1955 was in South America, 80 pounds per capita; while Central America (including Mexico and islands of the West Atlantic) consumed about 52 pounds per capita; Asia less than 7 pounds; Africa about 1.5 pounds; and Oceania, 27 pounds. In the United States and Canada combined consumption of bananas was 20 pounds per capita, and for Europe it was 9 pounds—an average of almost 13.5 pounds per capita for the two combined.

Table. Banana Production and Consumption (Stem-weight basis).

Area	Production Billion lbs.	Percent	Consumption Billion lbs.	Percent
Central America	6.0	23	3.0	12
South America	11.7	46	9.9	38
Asia	6.0	23	6.0	23
Africa	1.6	6	0.3	1
Europe	—	—	2.5	10
Oceania	0.4	2	0.4	2
North America	—	—	3.6	14
Total	25.6	100	25.6	100

Thus, up to the present time, it is fair to say that bananas have assumed as important a place in the average diet in the two major industrialized areas of the temperate zone as for the vast majority of the people living in what we regard as tropical areas. In India, for example, the apparent annual per capita consumption of bananas is only about 11 pounds. Since there is so little inter-country trade in bananas within the tropic zones, it follows that there is an extraordinarily high annual per capita consumption in some of the major producing countries, while in other tropical countries the consumption rate of bananas is very low.

For 1955, on the basis of *reported* production minus export figures, Brazilians consumed something like 150 pounds per capita; Costa Ricans, 350 pounds; Panamanians, 380 pounds; and Ecuadorians a fantastic 500 to 600 pounds per capita. Even allowing for apparent inflation in some of the reported production figures and for the known fact that there is considerable feeding of bananas to livestock in some of these countries, the human consumption of bananas where they are grown in profusion clearly reaches heroic proportions.

International Trade in Bananas

Of the estimated 25.7 billion pounds of bananas produced for commercial sale in 1955, something over 25 percent was exported in international trade. The breakdown by areas of export and import, is adapted from figures compiled by the U. S. Department of Agriculture. We have adjusted the North American import weights slightly upward upon the basis of evidence that the official estimates of the Department of Agriculture do not give sufficient allowance to the increases in average stem weight in recent years. This has the effect of raising the world total of 1955 imports from the Department's estimate of about 6.6 billion to 6.7 billion pounds.

It is probable that, on the export side, there is a comparable degree of understatement of weights for fruit going to North America but we have not attempted to make this adjustment in the data presented. The discrepancy is not large enough to alter substantially the proportionate shipments as reported by broad area. Reported import and export figures never are in exact balance because of inevitable inaccuracies in accounting and because, due to time consumed in ocean transport, there are always some shipments credited as exports at the end of one year that only show up as imports when delivered at the beginning of the following year.

On the export side, the data show that almost 80 percent of all reported banana exports in 1955 were shipped from Middle America, including Mexico and the islands of the West Atlantic, and from the three South American banana exporters—Ecuador, Colombia, and Brazil. Africa accounted for about 20 percent; Asia (entirely from Taiwan) for a bit more than 1 percent; and Oceania for less than 1 percent. It is clear, then, that Latin America is the dominant bananaexporting area of the world, and that the West Coast of Africa (in which the tabulation includes the Canary Islands) is the only other major area that plays a significant exporting role, and that a relatively minor one, in the overall picture.

On the import side, North America (United States and Canada) and Western Europe, neither of which produces any banana on a commercial scale, are the outlets for over 90 percent of world banana trade. Argentina, Chile, and Uruguay are the principal Latin American importers, with Brazil and Ecuador as sources of supply. The breakdown of world banana imports for 1955 is shown below.

Of the North American imports, about 92 percent went to the United States and a little more than 8 percent to Canada, most of the latter transshipped in bond from U. S. ports. Of the three South American importing countries, Argentina takes more than 80 percent of the total. African imports are divided among the Union of South Africa, Algeria, French Morocco, Southern Rhodesia, and Tunisia in the order of their importance as banana importers. Japan is the only listed Asian importer, with about 54 million pounds imported in 1955; and New Zealand the sole importer in Oceania at a slightly lower level than Japan.

Table. World Imports of Bananas, 1955.

Importing area	Billion lbs. imported	per cent of world imports
United States and Canada	3.60	53.8
Europe	2.47	36.9
South America	.43	6.4
Africa, Asia, Oceania	.20	2.9
Total	6.70	100.0

It now can be seen that this remarkable tropical fruit, originating in the Far East and for thousands of years rarely even listed in the learned chronicles of the West, has undergone a dramatic metamorphosis—largely within the last century. Almost in the manner of historic glacial drifts, it has crept from its original domicile in the East into western zones, until today of all bananas grown, the roots of 70 percent thrust down in the soil of the Western Hemisphere.

Table: Major World Fruit Exports.

Type of fruit	Export 1951- 53 average (1,000 metric tons)	per cent of freash fruit	per cent of total
Fresh fruit			
Bananas	2, 552	40.7	36.4
Orange and tangerines	2,060	32.8	29.4
Lemons and limes	261	4.2	3.7
Grapefruit	127	2.0	1.8
Apples, table	714	11.4	10.2
Pears, table	186	3.0	2.7
Grapes, table	218	3.5	3.1
Pineapples	154	2.4	2.2
Total	6,272	100.0	89.5
Dried fruit			
Dates	343		4.9
Raisins	285		4.0
Prunese	49		0.7
Dried figs	48		0.7
Other dried fruitf	12		0.2
Total dried fruit	737		10.5
Total all fruit	7.009		100.0

Another 5 percent grows along the African shores that rim the Atlantic. Considered in marketing terms, within the short span of the past 60 years this exotic fruit has burst the containing bounds of its central tropic environment, and has moved northward, and to a lesser extent to the south, to find a place in the fruit bowls of temperate zone tables from the Yukon to California, from Scandinavia to Austria and Italy, and from Chile to Uruguay.

Agriculturally, bananas are chained to the tropics. Commercially, about one-fourth of all that are produced find a market outside the temperature zones in which they thrive. And nine-tenths of this fourth find their way to the industrial heart of the western world-to North America and Western Europe, whose inhabitants a century ago scarcely knew what a banana was, except for the scholarly few who read the works of Pliny the Elder or of the Swedish botanist Linnaeus. An important part of our story is concerned with how this remarkable change came to be. It is the more remarkable in that it is a phenomenon that has occurred on a comparable scale with respect to no other delicately perishable, definitely tropical fruit.

The general picture of major world fruit exports, as the average exports in metric tons, 1951-53, is adapted from a study by Erik Mortenson, entitled "Trends in Production and Consumption of Fruit and Vegetables," in the September 1955 issue of the FAO publication, *Agricultural Economics and Statistics*. The table, which was evidently compiled on the basis of official FAO trade statistics, would appear to give a reasonable basis for comparing the relative magnitudes of the reported trade in the several fruits that bulk large in international commerce.

On this accounting, the trade in bananas for 1951-53 accounted for over 36 percent of all world commerce in fruit and for more than 40 percent of world trade in fresh fruits. Oranges and tangerines make up the only other fruit category that is of even proximate importance as a world trade item upon a weight basis, and the amount of bananas traded is about 25 percent larger. Citrus fruit, however, hardly classifies as a tropical fruit. Again using FAO figures, the 1954 record shows that almost two-thirds of commercially produced oranges and tangerines were grown in temperate zone areas, more than 50 percent in the United States and Western Europe alone.

Temperate zone countries furnish 70 percent of all exports. Over 85 percent of commercial grapefruit culture is restricted to the United States, and three-quarters of all commercially grown lemons and limes are produced in the United States, Europe, Japan, Argentina, and Uruguay. Aside from the fact that citrus fruit is far easier to ship than bananas, the striking fact is that all citrus shipments from *tropical areas* amount to only about one-quarter, by weight, of banana shipments, all of which originate in tropical areas.

Pineapples qualify as a tropical fruit crop, but the FOA figures show that only about 10 percent of total production is exported in world trade, and the great bulk of this was shipped in canned rather than fresh fruit form. Canned pineapples are included in the fresh fruit category upon the basis of the equivalent fresh fruit content. Dates are preponderantly a tropical area crop, and about one-fifth of reported production is exported internationally. But the big export is in dry rather than fresh fruit form and, in 1954, world trade in dates amounted to less than one-ninth that of bananas on a tonnage basis. Figs are the other entry on the list that in most people's minds would fall into the classification of tropical fruit. Actually, FAO figures assign considerably over half of commercial fig culture to Western Europe and the United States, and total fig exports, again with shipments preponderantly in dried form, account for less than I percent of world fruit trade. How did it come about that the banana (which is technically a herbaceous vegetable) should have become the one tropical "fruit" that has become a major item in the diet of North Americans and Europeans? Why the banana among the literally scores of fruits that are native to the tropics? Why not the mangosteen of which poet-gourmets have written in ecstatic terms, or the luscious mango, the peptic papaya, or the delicately flavoured naranjilla? It is certainly not because of special qualities that make the banana easy to grow or to transport.

We already have pointed out that the banana is singularly demanding with respect to the temperature, soil, precipitation, drainage, and wind conditions of its environment. The areas that can offer an even proximate optimum of all of these factors are relatively limited. As we shall see, bananas are subject to blights of devastating intensity for which it is tantalizingly difficult and formidably expensive to find adequate controls. This is virtually a seedless plant, with new cultivations established by planting large pseudobulbs or rhizomes, and there is no wood stalk to permit grafting.

Thus, there has been no success to date in developing, through the genetic approach, man-made strains that would combine the disease-resistant characteristics of certain varieties with the better qualities for handling and ripening inherent in others that are vulnerable to disease. In addition to disease blights, insect pests that destroy plant and fruit abound wherever bananas are grown.

The banana is a bulky fruit, shipped attached to a stem that accounts for about 7 to 8 percent of worthless weight, from which the fruit fingers protrude in a fashion that invites crushing and bruising under anything less than the tenderest handling. It must be cut in a green state, before full maturity—how long before depends upon the length of haulage to ultimate market—and the permissible time span between cutting and eating by the consumer is limited to not more than 21 to 25 days.

Yet, bananas regularly travel from 2,000 to 6,000 miles by water and up to 1,500 additional miles by rail and truck to reach their ultimate markets. Any accumulation of dust or gravel before or during transit, any roughness or cramping in stowing or carriage, even fingering by customers on the retailer's shelves shows up in marring discolouration of the delicate, golden skin of the ripened fruit and depreciates its marketability. Strict temperature controls have to be maintained from the time of loading aboard ship to delivery to retailer, and the latter is under strong compulsion to sell his fruit to customers within 24 hours after receipt.

Every stage, from plantation to retail sale, requires meticulous planning and coordination, upon a time schedule far more precise than pertains to any other major commodity of world trade. Supply in every market must be geared to a demand that, in turn, is affected by the availabilities of competitive fruits.

The cutting by maturity grade, the transport to shipping port, the stowing, the ocean carriage, the unloading, the sorting (by size, condition, and degree of ripeness), the shipment by truck or rail to distribution centres, the operations in the jobbers' ripening rooms, the trucking to retailers, and the sale to customers—every one of these stages is a tailored process with minimal tolerances for departure from a schedule that has to be calculated from its beginning. Even small divergences from the rigid timetable and handling requirements result in losses; whereas major departures would spell a total loss for the shipments involved.

If Captain Lorenzo Baker, Minor Keith, Thomas Hart, Andrew Preston, and even the imaginative Sam Zemurray had been able to visualize the complications that were to beset the production and marketing of bananas as a large-scale item in world trade, it is improbable that they would have had the fortitude to launch and expand the United Fruit Company. But if they had not made the start, it is probable that this thoroughly implausible commodity—implausible in the sense of its inherent lack of adaptability to the hazards of world marketing—would have bulked no larger in temperate zone food economies than do pineapples, or figs, or dates today.

Happily, for the North American and European consumer, and for the Latin American banana-producing economies as well, the vision of these pioneers was too limited to foresee the involvements or, at least, too sanguine to sense the formidable impediments that were to arise. So they built the foundations of today's giant trade, largely unaware of the whirling windmill blades before which even a Don Quixote might have quailed.

The Valuation of Market

The money value of bananas that enter into world trade, to the best of our knowledge, has had no systematic study. Values are commonly attributed to exports on an f.o.b. basis, and to imports on a c.i.f. basis. Such reporting is generally made by applying to the number of stems involved in either case a formula price that usually grossly understates the prices at which sales are actually made. Accordingly, the reported export valuation for a given country as shown in its annual trade account figures often represents 50 percent, or even less, of what it actually receives upon its adjusted accounts.

The International Monetary Fund, in its work of keeping track of international balance of payment flows, has been forced to make radical adjustments in reported banana trade figures to avoid untenably large distortions in the accounts of the major banana exporting countries. But the Fund only makes its adjustments for those countries in which banana exports account for a major portion of foreign exchange revenues.

The authors of this study, in attempting to put valuation figures on the world banana trade as a whole, have started with the North American market about which a good deal is known and which represented 54 percent of the world's total banana imports in 1955.

We have less complete data on the European marketing structure, which represented 37 percent of 1955 imports by weight, but there is sufficient evidence to indicate that an extrapolation of the American cost structure to that segment will not give any upward bias to the whole. The costs per pound of European bananas are consistently higher from original purchase in the producing areas on through the transportation and distribution chain. The remaining segment of world trade in bananas is too small to importantly distort the whole structure.

On a stem basis, the United Fruit Company accounted for 59 percent of North American imports, for 10 percent of European and for 28 percent of the world total. But on a weight basis—and bananas are sold by weight in markets of distribution—there are indications that the percentages of United Fruit's shipments are somewhat higher in each case, since the stem weight of the company's shipments generally runs above market averages.

Our specific data gleaned from the United Fruit Company's records furnished a far more precise base for the portion of the field that it covered than anything available in published studies. To this we added everything that could be furnished by public officials and private banana operators in the six countries in which we made detailed field surveys. As has been noted, these six, between them, produced about 60 percent of the bananas shipped in international trade in 1955. In sum, while our global estimates have been pieced together from these sources, and then blown up to represent the whole market as revealed by world trade and production statistics, we believe that our findings reasonably represent the general magnitudes of a field in which there has been no previous basis for any general appraisal.

It is our finding that, in 1955, North American consumers spent about $527 million for bananas—3,063 million pounds consumption weight; 3,604 million pounds stem weight—at an average retail price of 17.2¢ per pound. We know that Europeans paid a somewhat higher price per pound than Americans, and that other importing areas paid somewhat less. Taking the North American price as average, and knowing that this area accounted for 54 percent of all imports, we can estimate the world retail expenditure for *imported* bananas at approximately $976 million. We have seen earlier that about three-quarters of the reported world commercial production of bananas is consumed in the countries where they are produced. We have no overall reporting of what consumers paid for bananas in the producing countries, but we have some data on this for our six-country sample. In these countries, bananas sold on the local market for from one-fifth to one tenth of what they brought when sold as export stems.

It is certainly conservative to estimate that the totality of fruit grown for local markets throughout the world, although almost three times as great in quantity, has less than one-half and perhaps not more than a third the value of that committed to export markets. It is probably safe to estimate the total retail value of the 1955 world's commercial banana crop at between $1.3 billion and $1.5 billion.

We are able to make a reasonably proximate estimate of the value in 1955 of world banana exports f.o.b. vessels at ports of embarkation. On the basis of our North American consumer dollar analysis that follows, it will be seen that about 27¢ out of each such dollar spent on bananas represents the return actually realized by producing countries. Hence, of the estimated $976 million sales at retail, $263 million can be assigned as the share of the producing countries from their banana exports.

MANGO PROCESSING TECHNOLOGIES

Mangoes are processed at two stages of maturity. Green fruit is used to make chutney, pickles, curries and dehydrated products. The green fruit should be freshly picked from the tree. Fruit that is bruised, damaged, or that has prematurely fallen to the ground should not be used. Ripe mangoes are processed as canned and frozen slices, purée, juices, nectar and various dried products. Mangoes are processed into many other products for home use and by cottage industry.

The mango processing presents many problems as far as industrialization and market expansion is concerned. The trees are alternate bearing and the fruit has a short storage life; these factors make it difficult to process the crop in a continuous and regular way. The large number of varieties with their various attributes and deficiencies affects the quality and uniformity of processed products.

The lack of simple, reliable methods for determining the stage of maturity of varieties for processing also affects the quality of the finished products. Many of the processed products require peeled or peeled and sliced fruit. The lack of mechanised equipment for the peeling of ripe mangoes is a serious bottleneck for increasing the production of these products.

A significant problem in developing mechanised equipment is the large number of varieties available and their different sizes and shapes. The cost of processed mango products is also too expensive for the general population in the areas where most mangoes are grown. There is, however, a considerable export potential to developed countries but in these countries the processed mango products must compete with established processed fruits of high quality and relatively low cost.

GREEN MANGO PROCESSING

Pickles

The optimum stage of maturity should be determined for each variety used to make pickles. There are two classifications of pickles - salt pickles and oil pickles. They are processed from whole and sliced fruit with and without stones. Salt is used in most pickles.

The many kinds of pickles vary mainly in the proportions and kinds of spices used in their preparation. One basic recipe for the study of the preparation and storage of pickles in oil is as follows:

Mango pieces	250 g	Tumeric powder	2 to 4 g
Salt	60 g	Fenugreek seeds	2 to 4 g
Mustard powder	30 g	Bengal gram seeds	2 to 4 g
Chili powder	20 g	Gingelly oil	20 to 30 g

The ingredients are mixed together and filled into wide-mouthed bottles of 0.5 kg capacity. Three days later the contents are thoroughly mixed and refilled into the bottles. Extra oil is added to form a 1-2 cm layer over the pickles.

Chutney

The product is prepared from peeled, sliced or grated unripe or semi-ripe fruit by cooking the shredded fruit with salt over medium heat for 5 to 7 minutes, mixed and then sugar, spices and vinegar are added. Cook over moderate heat until the product resembles a thick purée, add remaining ingredients and simmer another 5 min. Cool and preserve in sterilised jars.

Spices usually include cumin seeds, ground cloves, cinnamon, chili powder, ginger and nutmeg. Other ingredients such as dried fruits, onions, garlic and nuts may be added.

Drying/dehydration. immature fruit is peeled and sliced for sun-drying. The dried mango slices can be powdered to make a product called amchoo. The use of blanching, sulphuring and mechanical dehydration gives a product with better colour, nutrition, storability and fewer microbiological problems.

RIPE MANGO PROCESSING

Puree

Mangoes are processed into purée for re-manufacturing into products such as nectar, juice, squash, jam, jelly and dehydrated products. The purée can be preserved by chemical means, or frozen, or canned and stored in barrels. This allows a supply of raw materials during the remainder of the year when fresh mangoes are not available. It also provides a more economical means of storage compared with the cost of storing the finished products, except for those which are dehydrated, and provides for more orderly processing during peak availability of fresh mangoes.

Mangoes can be processed into purée from whole or peeled fruit. Because of the time and cost of peeling, this step is best avoided but with some varieties it may be necessary to avoid off-flavours which may be present in the skin. The most common way of removing the skin is hand-peeling with knives but this is time-consuming and expensive. Steam and lyepeeling have been accomplished for some varieties.

Several methods have been devised to remove the pulp from the fresh ripe mangoes without hand-peeling. A simplified method is as follows: the whole mangoes were exposed to atmospheric steam for 2 to 2 1/2 min in a loosely covered chamber, then transferred to a stainless steel tank.

The steam-softened skins allowed the fruit to be pulped by a power stirrer fitted with a saw-toothed propeller blade mounted 12.7 to 15.2 cm below a regular propeller blade. The pulp is removed from the seeds by a continuous centrifuge designed for use in passion fruit extraction. The pulp material is then passed through a paddle pulper fitted with a 0.084 cm screen to remove fibre and small pieces of pulp.

Mango purée can be frozen, canned or stored in barrels for later processing. In all these cases, heating is necessary to preserve the quality of the mango purée. In one process, purée is pumped through a plate heat exchanger and heated to 90°C for 1 min and cooled to 35° C before being filled into 30 lb tins with polyethylene liners and frozen at -23.50 C.

In an other process, pulp is acidified to pH 3.5, pasteurized at 90°C, and hot-filled into 6 kg high-density bulk polyethylene containers that have been previously sterilised with boiling water. The containers are then sealed and cooled in water. This makes it possible to avoid the high cost of cans.

Wooden barrels may be used to store mango pulp in the manufacture of jams and squashes. The pulp is acidified with 0.5 to 1.0per cent citric acid, heated to boiling, cooled, and SO2 is added at a level of 1000 to 1500 ppm in the pulp. The pulp is then filled into barrels for future use.

Slices

Mango slices can be preserved by canning or freezing, and recent studies have shown the feasibility of pasteurized-refrigerated and dehydro-canned slices. The quality of the processed product in all of these procedures will be dependent upon selection of a suitable variety along with good processing procedures. Thermal process canning of mango slices in syrup is the most widely used preservation method.

Beverages

The commercial beverages are juice, nectar and squash. Mango nectar and juice contain mango purée, sugar, water and citric acid in various proportions depending on local taste, government standards of identity, pH control, and fruit composition of the variety used. Mango squash in addition to the above may contain SO2 or sodium benzoate as a preservative. Other food grade additives such as ascorbic acid, food colouring, or thickeners may be used in mango beverages. A short description of finished products found in literature is as follows:

- Mango juice: prepared by mixing equal quantities of pulp (purée) and water together and adjusting the total soluble solids (TSS) and acidity to taste (12 to 15per cent TSS and 0.4 to 0.5per cent acidity as citric acid);
- Mango nectar containing 25per cent purée can be prepared using the following procedure.

Table. Brix of Puree.

Nectar components	15°	17°	20°
Purée	100	100	100
Sugar	45	43	40
Water	255	257	260

Commercial processing conditions may require the use of a preservative.

The pH is adjusted to approximately 3.5 by adding citric acid as a 50per cent solution.

The time of heat processing will vary with filling temperature, can size and viscosity of the juice or nectar.

Mango squash may be prepared according to flow-sheet described below; the finished product may contain 25per cent juice, 45per cent TSS and 1.2 to 1.5per cent acidity and may be preserved with sulphur dioxide (350ppm) or sodium benzoate (1000 ppm) in glass bottles. Mango squash simplified flow-sheet.

Ingredients

Mango pulp	900	900
Sugar	900	1100
Citric acid	18	15
Water	900	900

Mangoes are washed, stored, peeled with stainless steel knives. The pulp is prepared by using a pulper with fine sieve (0.025-in); Sugar is mixed with water and citric acid = syrup; The pulp is added to the syrup and mixed well; The mixture is strained trough cloth; The squash is heated at 85° C and bottles are filled and closed.

For additional heat treatment bottles may need to be maintained at a product temperature of 80°C for 30 minutes if the product is to be processed without preservatives. The bottles are then left to cool in water and stored at room temperature.

Two negative points must be avoided: presence of air bubbles (which is a source of quick deterioration) and separation of squash solids (giving an undesirable appearance). The means to avoid these two phenomena are described in the fruit juices section.

A type of "squash type" beverage may also be manufactured with '/a pulp + '/a water + i/a sugar and pH adjusted to 3.7 by addition of citric acid. Using different sieve sizes affects the quality and reduces air bubbles to a certain extent but homogenisation and de-aeration of purée or squash seem to be important in order to avoid separation and air bubbles. The squash quality is evaluated on the basis of the following characteristics: pH, titrable acidity, soluble solids, ascorbic acid (by 2,6 dichlorophenol indophenol method), specific gravity.

Dried/Dehydrated

Ripe mangoes are dried in the form of pieces, powders, and flakes. Drying procedures such as sun-drying, tunnel dehydration, vacuum-drying, osmotic dehydration may be used. Packaged and stored properly, dried mango products are stable and nutritious.

One described process involves as pre-treatment dipping mango slices for 18 hr (ratio 1:1) in a solution containing 40° Brix sugar, 3000 ppm SO_2, 0.2per cent ascorbic acid and 1per cent citric acid; this method is described as producing the best dehydrated product. Drying is described using an electric cabinet through flow dryer operated at 60° C. The product showed no browning after 1 year of storage.

Drum-drying of mango purée is described as an efficient, economical process for producing dried mango powder and flakes. Its major drawback is that the severity of heat preprocessing can produce undesirable cooked flavours and aromas in the dried product. The drum-dried products are also extremely hydroscopic and the use of in-package desiccant is recommended during storage.

Canning

This preservation technology is described in various technological flow-sheets in this bulletin. Mango bar or "fruit leather" is presented in various flow-sheets.

MANGO MALFORMATION

It is the most important malady of mango and was first reported by Burn (1910). It causes the losses from 50-80 per cent. It is also reported to be due to other disorders like physiological and acarological. Indications of its virus nature has also been postulated. The maximum incidence of the disease has been recorded in the north, north-west and north-east parts of the country. The incidence is, however, less in western and southern India. There are two types of symptoms, namely, floral malformation and witches broom or bunchy top or vegetative malformation with a proliferation of infected tissue. The flowering panicles turn into a compact mass of flowers.

This compact mass is very hard and not soft like normal panicle. Individual flower is greatly enlarged and has a large disc. The percentage of bisexual flowers in malformed panicles is very low. In bunchy top, compact leaves are formed in a bunch at the apex of shoot or in the leaf axil.

A similar bunch consisting of small rudiments crowded together on short shootlets is seen in vegetative malformation in which the growth of shootlet is arrested. Vegetative malformation is more pronounced in young seedling and seedling trees. The malformed heads dry up in black masses and persist on the trees for a long time. The malformed inflorescences contain more of endogenous cytokinins than healthy ones (Nicholson and van Staden 1988).

The variety Zebda has been reported to be resistant in Egypt (El Ghandour *et al.* 1979). Considerable incidence reduction has been reported by spraying 100-200 ppm NAA during October.

Use of disease free planting material and prophylactic spray of insecticides and fungicides can keep the orchards healthy. In areas with less than 5-10 per cent infection, pruning of diseased plants should be made compulsory; pruning of diseased parts along the basal 15-20 cm apparently healthy portions. This is followed by the spraying of Bavistin (0.1per cent) or Captaf (0.2per cent). Partial control of malformation has been achieved by spraying mangiferin-Zn2 and mangiferin-Cu2+chelates.

Stem End Rot (Botryodiplodia theobromae Pat.)

It is an important disease of ripe mango. This fungus also causes die-back symptoms in orchards. It is known to occur in India, Myanmar, Sri Lanka, the Philippines, Mauritius and the U.S.A. In India, it is recorded from all the mango growing areas. The disease is characterized as the dark epicarp around the base of the pedicel in the initial stage. In the next few hours, the affected area enlarges to form a circular, black patch which in humid atmosphere extends rapidly and turns the whole fruit completely black within two or three days. The pulp of the diseased fruits become brown and somewhat softer. The diseased fruits showed increased phosphorus and reduced ascorbic acid contents. Pectolytic enzyme activity was also lowered in such fruits. The disease is carried from the dead twigs and bark of the trees harbouring the pathogen and with the onset of rains the orchards get contaminated.

Dipping of mangoes in 6 per cent borax solution at 43°C for 3 minutes reduces the incidence. Fruit dipping in Aureofungin and wrapping was also found effective. Pathak (1980) suggested to harvest mangoes on clear dry day. Hot water treatment at 53°C for 10 minutes and at 51 to 55°C for 15 minutes have been recommended elsewhere. Care should be taken to prevent snapping-off of the pedicels. Injury should be avoided to fruits at all stages of handling. Spalding (1982) in his experiments found thiobendazole and thiophenate methyl to be the most effective ones. Recently, Rawal and Ullasa (1988) have reported that Bavistin (0.1per cent) or Topsin-M (0.1per cent) or chlorothalonil (0.2per cent) spraying in the field before harvesting give very effective control.

Die-back (Botryodiplodia thiobromae Pat)

It is a descructive disease of mango and is known to occur in India and other mango growing countries. The disease is manifested by discolouration and darkening of the bark some distance from the tip of the twigs. The dark area advances and young green leaves start withering first at the base and then extending outwards along the vein. The affected leaves turn brown and the margins roll upward. At this stage, the twig or branch dies, leaves shrivel and fall, and this may be accompanied by exudation of gum. The infected twigs show internal discolouration. Brown streaks of vascular tissues is seen on splitting the twigs lengthwise along the long axis. In early stages, epidermal and sub-epidermal cells of twigs appear slightly shrivelled. The areas of cambium and phloem show brown discolouration and yellow gum like exudate flows out of the cells. The disease is known to be enhanced by a beetle, Xyleborus affinis (Batista 1947). Relative humidity above 80per cent and temperature of 25 to 31.5°C, and rains enhance disease development. The die-back is also caused due to Colletotrichum gloeosporioides. Both these pathogens also attack fruits which get spoiled during ripening.

For effective control of the disease, pruning and destruction of infected twigs is the foremost practice. Spraying the trees periodically with copper oxychloride sulphate is recommended by Alvarez and Lopez (1971). Pasting of trees with a mixture of oil and 5per cent phenol was found effective. Sprays of carbendazim (0.1per cent) or methyl thiophenate (0.1per cent) or chlorothalonil (0.2per cent) at fortnightly interval during rainy season is important. Sometimes short hole borers also predispose the trees to infection and hence proper insecticides are also to be sprayed. Healthy twigs should be selected for grafting of seedling during propagation.

Leaf Blight (Macrophomina mangiferae Hingorani and Sharma)

It was first reported from Brazil to be caused by Macrophomina sp. by Lacca (1922). In India, it was reported by Patel *et al.* (1949). Later, it has been reported by various workers from Delhi, U.P. and Gujarat. The other pathogens causing the disease are grey blight (Pestalotiopsis mangiferae), phoma blight (Phoma glomerate), and twig blight and fruit rot (Phoma sp.). The disease usually appears as yellowish pin head like spots on leaves and twigs of the affected plants. These gradually enlarge discolouring the surrounding tissues which first become brown and then dark brown with slightly raised and brown purplish margins and later ash coloured due to the appearance of pycnidia.

Spots are round to start with but later become oval or irregular in size depending upon environmental conditions such as humidity and temperature. Infection is mainly observed on leaves and rarely on stems. On stem, the lesions are elliptic which later girdle the stem at the point of infection. On fruits, water soaked circular lesion are produced which enlarge rapidly and cause rotting.

The pathogen can survive for more than a year on the leaves of mango. The varieties, namely, Khandeshi, Borasio and Asal Damido are reported to be resistant. Field sanitation is by collecting and destruction of diseased parts. Spraying of Burgandy mixture, Perenox, lime sulphur and dithane have been recommended by Hingorani *et al.* (1960).

Red Rust (Cephaleuros mycoides Karst)

This or red spot is a common algal disease on the mango in the tarai and in the other humid regions of India. Its occurrence has been reported from Bihar, Karnataka and Uttar Pradesh.

The alga attacks the foliage, bark and twigs of the host plant and is both parasitic and epiphytic in nature. In serious infections, the bark becomes thickened, the twigs get enlarged and remain stunted and the foliage becomes sparse and finally dries up. Initially the spots are greenish-grey in colour and velvety in texture, but later on the surface bears reddish-brown appearance. Algal spot is circular to irregular in shape, slightly elevated and with usually 2 mm in diameter, though in some cases it may be as much as 1 cm. Bordeaux mixture (6:6:100), Supramar, Fytolan, Blitox-50 or lime sulphur are recommended.

Black Spot (Xanthomonas compestris pv mangiferae indicae)

It is one of the serious diseases and is reported from many countries. In India, it is reported from Maharashtra, Delhi, Uttar Pradesh, Tamil Nadu and Karnataka. It affects leaves, petioles, fruits and tender stems. Numerous small angular water soaked lesions appear in groups towards the tip of the leaf blade. These are first light yellowish but later turn dark brown to black and get surrounded by a distinct halo. Several lesions are often found to coalesce forming large necrotic spots. On young fruits, water-soaked lesions develop which also turn dark brown to black. Infected fruits may show skin cracking and badly affected ones dry prematurely. Infection by wounding produces diseased spots on tender stems, petioles and stalk bearing fruits. These spots turn into deep longitudinal scars accompanied by appreciable amount of bacterial gum-like ooze. The incubation period on leaves varies from 2-14 days. In India, the disease remains dormant during November to March due to low temperature (11.8-

22°C) and the leaf infection is considerably reduced by the fall of the infected leaves. Kent mangoes showed close relationship of rainy season with incidence of bacterial leaf spot. One day's rainfall did not affect the disease appearance, whereas four days of rainfall led to the 36.9per cent disease incidence.

Twig canker persists and initiates fruit infection. The disease spread is rapid during the rains and becomes severe in July-August. Five application of Bordeaux mixture (4:4:50) with spreader was recommended by Wager (1937), whereas four sprays of 6:6:50 Bordeaux mixture was found to be effective by Marloth (1947). Sprays with fungicides such as copper containing materials and agrimycin have also been advocated. Shekhawat and Patel (1975) recommended orchard sanitation by way of removal of infected materials and seedling treatment as preventive measures. Agrimycin-100 proved best in arresting the disease development. Plantomycin (200 ppm) followed by agrimycin + bavistin (1000 ppm) was reported to be the best against this disease. Streptomycin sulphte followed by aureofungin have been recommended by Prakash and Raooff (1985) to control bacterial disease of mango.

Black Mould Rot (Aspergillus niger Van Teigh)

The affected fruits show yellowing of base and development of irregular, hazy, greyish spots and coalesce into dark brown or black lesions. The mesocarp of the rotted areas becomes depressed and soft. The diseased fruits show rapid decrease in ascorbic acid content. The pathogen infects injured fruits. Totapuri variety is highly susceptible. A fruit dip treatment with benlate at 1500 ppm can control the rot.

Soft Rot (Rhizopus arrhizus Fisc.)

This pathogen alone is reported to be responsible for the decay of 6.3 per cent fruits in Delhi markets. The pathogen incites a typical soft rot. Development of disease is slow and less intensive at low temperature (7-8°C). The disease develops rapidly between 20 and 40°C. Thakur and Chenulu (1970) found that 2-aminothiozole with 2-aminopyridine each at 7 per cent concentration protected mango fruits for 20 days.

Some of the other post-harvest diseases and pathogens are Alternaria rot (Alternaria tenuissima (Neer. Ex. fr.) Wiltshire), Sclerotium rot (Sclerotium rolfsii Sacc.), dry rot (Boothiella tetraspora Lodhi and Mirza), brown rot (Pestalotiopsis mangiferae P. Henn. and P. glandiola (Cast) Stey), fruit rot (Phomopsis mangiferae Ahmed) and black spot, Actinodichium jenkinsii.

Black Tip

This disease is peculiar to India and does not occur in any other country. It has been reported to occur in regions of West Bengal, Uttar Pradesh, Bihar and Punjab in orchards located in close proximity of a brick kiln. South India is free from this disease. The symptoms manifest as small etiolated area at the distal end of the fruit, which gradually spreads, turns nearly black and covers the tip completely. Before the etiolation is complete, isolated greyish spots appear which become dark brown, enlarge and coalesce into a continuous necrotic area. The brown discolouration spreads to the neighbouring parenchyma and the deposits also appear in the ducts. The browning and deposits gradually spread throughout the mesocarp and the affected cells distintegrate and coalesce into a dead tissue. In severe cases, the necrosis extends to the endocarp. The practical control measure lies in keeping the brick kilns away from the mango orchards. The use of telescoping chimney 12-15 m high has also been recommended. Spraying of borax (0.6per cent) at 10-14 day intervals has been recommended.

Soft Nose

The problem was first observed as physiological breakdown in cvs. Kent and Haden by orchardists in 1950 in Florida. The soft nose is characterized by hallowing of the green skin in the areas between the apex and the stigmal point. A lack of firmness in diseased fruits can also be felt with experience. Upon cutting the flesh on the ventral side towards the apex of some soft nose fruits appear to be over ripe, while on shoulder and dorsal side it is unripe. The over ripe flesh surrounds a mass of yellowish to brown tissue which is firmer than the surrounding affected tissue and is bitter in taste.

Young (1957) concluded that the disorder is not due to the microorganisms. However, Young *et al.* (1963) reported the loss of only few percent in fruits harvested from trees grown on calcareous rock soil, but on acidic soil, sandy soil it is common for 15-20 per cent incidence. According to Young *et al.* (1963), soft nose disorder increased with N fertilization but was reduced to greater extent by increasing Ca level of the soil. Nitrogen fertilizer in sandy acidic soil increased the incidence whereas in calcareous soil there was no effect. To overcome this disorder, the high N level of soil should not be allowed in the sandy or acidic soil and proper Ca level should also be maintained in soil. Early picking is also suggested. The other disorders are little leaf caused by zinc deficiency, tip burn due to moisture stress and salt accumulation in soil and leaf scorch due to accumulation of chloride ion in leaf.

Phanerogamic Parasites (Dendrophthoe (Loranthus) viscum)

The mango trees are severely affected by this parasite particularly in neglected orchards. They are commonly present on the trunk or branches of the trees. The foliage of the infected tree is sparse, reduced in size and its bearing capacity and quality of fruits is considerably reduced. The fruits of this parasite are baccate with viscous mesocarp and attract the birds by its attractive colour. The seeds of this parasite are spread through bird and animal droppings.

Such seeds, germinate and produce haustoria. Sometime tumour like swellings are observed on the affected tissue. There are many other species of parasites which also attack mango trees. All epiphytic plants can be removed easily in early stages before their aerial roots penetrate the ground and surround the host trunk.

Some more diseases reported by various workers attacking the mango crop in the field are leaf spots due to Rhizoctonia bataticola (Taub) Butt., Cercospora mangiferae indica Munjal Lal and Chona, Alternaria alternata Fr. and Keissler, Phoma sorghicola (Sacc.) Boerema. Dorem and Venkast; Scab (Elsinoe mangiferae Bit and Jen; black band (Rhinocladium corticolum Massee); pink disease (Pellicularia solmonicolour (Berk and Br.) Dastur); root rot and damping-off (Rhizoctonia solani Kuhn) and wilt due to Verticillium albo-atrum Reinke and Berth and Fusarium solani (Mart.) Sacc. There are several more fungi causing different diseases but the above are the most important among them.

RECENT TRENDS IN FRUIT AND VEGETABLE PROCESSING

New Products

The number and variety of fruit and vegetable products available to the consumer has increased substantially in recent years. The fruit and vegetable industry has undoubtedly benefited from the increased recognition and emphasis on the importance of these products in a healthy diet.

Traditional processing and preservation technologies such as heating, freezing and drying together with the more recent commercial introduction of chilling continue to provide the consumer with increased choice. This has been achieved by new heating (*e.g.* UHT, microwave, ohmic) and freezing (*e.g.* cryogenic) techniques combined with new packaging materials and technologies (*e.g.* aseptic, modified atmosphere packaging).

The overall trend in new fruit and vegetable products is "added value", thus providing increased convenience to the consumer by having much greater variety of ready prepared fruit and vegetable products. These may comprise complete meals or individual components. The suitability of products and packages for microwave re-heating has been an important factor with respect to added convenience.

Fruit and vegetable product trends

Heat processed products

- Canned fruits and vegetables
  - Combination of vegetables in sauces and vegetable recipe dishes. Exotic fruits.
- Glass packed fruits and vegetables
  - "Condiverde"/"antipasti" products based on vegetables in oil.
  - High quality fruit packs.

- Retortable plastics
  - Basic vegetables or vegetable meals
  - Fruit in jelly
- Aseptic cartons
  - Ready made jelly
- Rosti meals
  - Potato based products in retort pouches
- Fruit juices
  - New combinations of juices and freshly squeezed products
- Crisps
  - Thick and crunch skin-on crisps. Kettle or pan fried chips. Lower fat crisps.

Table. Numbers of New Fruit and Vegetable Products.

	1990	1991	1992 Jan-June
Frozen vegetable products	66	95	21
Chilled vegetable products	76	81	78
Heat processed vegetables	51	60	38
Heat processed fruits	13	14	5
Fruit juices and drinks	73	83	46
Potato crisps	32	33	16

New product development in the fruit and vegetable sector is most important in meeting the continued challenge of providing the consumer with choice and high quality products.

A Fresh Look at Dried Fruit

New fruit varieties and advance in drying technologies are putting a fresh twist on dried fruit applications. Fruits that have been introduced to the drying process include cranberries, blueberries, cherries, apples, raspberries and strawberries - not to mention the traditional mainstays of raisins, dates, apricots, peaches, prunes and figs.

Perceived as a "value-added" ingredient, dried fruit adds flavour, colour, texture and diversity with little alteration to an existing formula. The growing interest in ethnic cuisines in U.S.A. and the change to a more healthy way of eating, has also moved dried fruit considerably closer to the mainstream.

Found primarily in the baking industry, dried fruit is coming into its own in various food products, including entrees, side dishes and condiments. Compotes, chutneys, rice and grain dishes, stuffings, sauces, breads, muffins, cookies, deserts, cereals and snacks are all food categories encompassing dried fruit.

Since some dried fruit is sugar infused (osmotic drying), food processors can decrease the amount of sugar in formula - this is especially the case in baked products. Processors are making adjustments in moisture content of the dried fruit so that a varied range is available for different applications. An added bonus is dried fruits' shelf stability (a shelf life of at least 12 months). Dried fruit is more widely available in different forms, including whole dried, cut, diced and powders.

Citric Acid and its Use in Fruit and Vegetable Processing

Citric acid may be considered as "Nature's acidulant". It is found in the tissues of almost all plants and animals, as well as many yeasts and moulds. Commercially citric acid is manufactured under controlled fermentation conditions that produce citric acid as a metabolic intermediate from naturally-occurring yeasts, moulds and nutrients. The recovery process of citric acid is through crystallization from aqueous solutions.

Citric acid is widely used in carbonated and still beverages, to impart a fresh-fruit "tanginess". Citric acid provides uniform acidity, and its light fruity character blends well and enhances fruit juices, resulting in improved palatability. The amount of citric acid used depends on the particular desired flavour (*e.g.*, High-acid: lemonade; Medium-acid: orange, punch, cherry; Low-acid: strawberry, black cherry, grape).

Sodium citrate is often added to beverages to mellow the tart taste of high acid concentrations. It provides a cool, distinctive smooth taste and masks any bitter aftertaste of artificial sweeteners. In addition, it serves as a buffer to stabilise the pH at the desired level. The high water solubility of citric acid (181 g/100 ml) makes it an ideal additive for fountain fruit syrups and beverages concentrates as a flavour enhancer and microbial growth inhibitor (preferably at pH < 4.6).

In processed fruits and vegetables, citric acid performs the following functions:

- It reduces heat-processing requirements by lowering pH: inhibition of microbial growth is a function of pH and heat treatment. Higher heat exposure and lower pH result in greater inhibition. Thus the use of citric acid to bring pH below 4.6 can reduce the heating requirements. In canned vegetables, citric acid usage is greatest in tomatoes, onions and pimentos. For tomato packs, the National Canners Association recommends a pH of 4.1 to 4.3. In general, 0. 1per cent citric acid will reduce the pH of canned tomatoes by 0.2 pH units.
- Optimise flavour: citric acid is added to canned fruits to provide for adequate tartness. Recommended usage level is generally less then 0.15per cent.
- Supplement antioxidant potential: citric acid is used in conjunction with antioxidants such as ascorbic and erythorbic acids, to inhibit colour and flavour deterioration caused by metal-catalysed enzymatic oxidation. Recommended usage levels are generally 0.1per cent to 0.3per cent with the antioxidant at 100 to 200 ppm.

- Inactivate undesirable enzymes: oxidative browning in most fruits and vegetables is catalysed by the naturally present polyphenol oxidase. The enzymatic activity is strongly dependent on pH.
- Addition of citric acid to reduce pH below 3 will result in inactivation of this enzyme and prevention of browning reactions.

Cherry and Apricot Oils are Safe for Food Use

The oils obtained by cold pressing the kernels of the cherry (Prunus cerasus) and the apricot (Prunus armeniaca) have been declared acceptable for food use by the UK Ministry of Agriculture, Fisheries and Food's Advisory Committee on Novel Foods and Processes (ACNFP) by June 1993.

In its assessments of the safety in use of the cherry and apricot kernel oils, the ACNFP consider specifications that included data on fatty acid composition, the presence of natural antioxidants and the content of cyanide, mycotoxins and heavy metals.

The Committee says that it gave particular consideration to the possible presence in the oils of the cyanogenic glucoside amygdalin, from which cyanide is released by enzymic action when the kernels of cherry and apricot are crushed. Amygdalin was found to be absent from the cherry and apricot kernel oils.

The ACNFP is satisfied that there are no food safety reasons why the use of cherry and apricot kernel should not be acceptable provided there is compliance with the specifications shown in Table given below.

Table. Specification of Purity for Cherry and Apricot Kernel Oils as Determined by UK ACNFP.

	Cherry	Apricot
Contaminants limits:		
Heavy metals (total)	0.5 mg/kg	0.5 mg/kg
Aflatoxins (total)	4.0 g/kg	4.0 g/kg
Cyanide	0.15 mg/kg	0.15 mg/kg
Pesticide residues	0.01 mg/kg	0.01 mg/kg
Tocopherols:		
Alpha/delta/gamma (mg/kg)	356-886	569-899

The oils are obtained by the mechanical mincing and cold pressing of kernels extracted from cleaned cherry or apricot stones. After filtering, the oils are stored and are to be sold in a raw, unrefined state. The cherry and apricot kernel oils are high unsaturated and are expected to be used as speciality oils for salad dressings, baking and shallow frying applications.

The Use of Fruit Juices in Confectionery Products

During the last decade, the concept of fruit juices has gained immensely on consumer popularity. The majority of new non-alcoholic and alcoholic fruit drink products were a combination of syrups, fruit juices and flavours.

The confectionery industry followed suit and new products incorporated fruit juices as part of their confectionery formulations and processes. Fruit juice concentrates of high solids are often used instead of normal or single-fold juices.

Juice concentrates are made of pure fruit juices. The process starts with pressing fruits and obtaining pure fruit juice; this is stabilised by heat treatment which inactivates enzymes and micro-organisms. The next processing step is concentration under vacuum up to 40-65° Brix or 4-7 fold. The concentrates are then blended for standardisation and stored.

These fruit juice concentrates are often further stabilised by the addition of sodium benzoate and potassium sorbate and are usually stored away from light and are refrigerated or frozen.

Depectinised fruit juices are also used to prevent foaming in confectionery processes and are essential for use in clear beverage products. Fruit juice concentrates which are depectinised, and have added preservatives are called stabilised, clarified, fruit juice concentrates.

Fruit juices are used in confectionery products in conjunction with natural and artificial flavours which provides intense flavour impact and are cost-effective for a confectionery product.

The traditional concern in using fruit juice concentrates in confectionery applications has been the effect of the natural acids on the finished product, particularly the formation of invert sugar during processing.

This is a logical concern since concentrates contain differing amounts and types of acids. For example: apple, cherry, strawberry and other berries contain primarily malic acid. Grapes mainly contain tartaric acid. Cranberry is high in quinic acid. Citrus fruits and pineapple contain differing amounts of citric acid. The concentrates, when used, are normally buffered to a pH of 5-7 with sodium hydroxide.

In formulating products with fruit juice concentrates, the solids of the concentrate are considered as mostly reducing sugars and a reduction in corn syrup is made to compensate for equivalent amount of reducing sugar being added in the concentrate.

The exact replacement can be determined by measuring the D.E. of the concentrate to be added. In formulations when small amounts of concentrate are used (less than 1per cent), no adjustment is made since the reducing sugar contribution of the concentrate is not significant.

Fruit juice concentrates can also be used to provide a source of natural colour, in particular red colour. Grape, raspberry, cherry, strawberry and cranberry concentrates in small amounts are very effective in colouring cream centres.

The inclusion of fruit juices in confectionery products is now left up to the imagination of the manufacturer. These products must, of course, hold up to the standards of flavour integrity, and product excellence, during the shelf-life of these products.

VEGETABLE SPECIFIC PROCESSING TECHNOLOGIES

VEGETABLES VARIETIES

Vegetable processors must appreciate the substantial differences that varieties of a given vegetable will possess. In addition to variety and genetic strain differences with respect to weather, insect and disease resistance, varieties of a given vegetable will differ in size, shape, time of maturity, and resistance to physical damage.

Varietal differences then further extend into warehouse storage stability, and suitability for such processing methods as canning, freezing, pickling or drying. A variety of peas that is suitable for canning may be quite unsatisfactory for freezing and varieties of potatoes that are preferred for freezing may be less satisfactory for drying or potato chip manufacture.

This should be expected since different varieties of a given vegetable will vary somewhat in chemical composition, cellular structure and biological activity of their enzyme system.

Harvesting and pre-processing

When vegetables are maturing in the field they are changing from day to day. There is a time when the vegetable will be at peak quality from the standpoint of colour, texture and flavour. This peak quality is quick in passing and may last only a day. Harvesting and processing of several vegetables, including tomatoes, corn and peas are rigidly scheduled to capture this peak quality.

After the vegetable is harvested it may quickly pass beyond the peak quality condition. This is independent of microbiological spoilage; these main deteriorations are related to:

- Loss of sugars due to their consumption during respiration or their conversion to starch; losses are slower under refrigeration but there is still a great change in vegetable sweetness and freshness of flavour within 2 or 3 days;
- Production of heat when large stockpiles of vegetables are transported or held prior to processing.

At room temperature some vegetables will liberate heat at a rate of 127,000 kJ/ton/day; this is enough for each ton of vegetables to melt 363 kg of ice per day. Since the heat further deteriorates the vegetables and speeds microorganisms growth, the harvested vegetables must be cooled if not processed immediately. But cooling only slows down the rate of deterioration, it does not prevent it, and vegetables differ in their resistance to cold storage. Each vegetable has its optimum cold storage temperature which may be between about 0-100 C (32-500 F).

- The continual loss of water by harvested vegetables due to transpiration, respiration and physical drying of cut surfaces results in wilting of leafy vegetables, loss of plumpness of fleshy vegetables and loss of weight of both.

Moisture loss cannot be completely and effectively prevented by hermetic packaging. This was tried with plastic bags for fresh vegetables in supermarkets but the bags became moisture fogged, and deterioration of certain vegetables was accelerated because of buildup of CO2 and decrease of oxygen in the package. It therefore is common to perforate such bags to prevent these defects as well as to minimise high humidity in the package which would encourage microbial growth.

Shippers of fresh vegetables and vegetable processors, whether they can, freeze, dehydrate, or manufacture soups or ketchup, appreciate the instability and perishability of vegetables and so do everything they can to minimise delays in processing of the fresh product. In many processing plants it is common practice to process vegetables immediately as they are received from the field.

To ensure a steady supply of top quality produce during the harvesting period the large food processors will employ trained field men; they will advise on growing practices and on spacing of plantings so that vegetables will mature and can be harvested in rhythm with the processing plant capabilities. This minimises stockpiling and need for storage.

Cooling of vegetables in the field is common practice in some areas. Liquid nitrogen-cooled trucks may next provide transportation of fresh produce to the processing plant or directly to market. Upon arrival of vegetables at the processing centre the usual operations of cleaning, grading, peeling, cutting and the like are performed using a moderate amount of equipment but a good deal of hand labour also still remains.

Reception

This covers qualitative and quantitative control of delivered vegetables. The organoleptic control and the evaluation of the sanitary state, even if they are very important steps in vegetables' characteristics assessment, cannot establish their technological value.

On the other hand, labouratory controls do not precisely establish their technological properties because of the difficulty in putting into showing some deterioration when using rapid control methods.

One correct method of vegetable quality appraisal is their overall evaluation based on the whole complex of data that can be obtained by combining an extensive organoleptic evaluation with simple analysis that can be performed rapidly in plant labouratory. These analysis can be:

- Refractometric extract (tomatoes, fruit, etc.);
- Specific weight (potatoes, peas, etc.);
- Consistency (measured with tenderometres, penetrometres, etc.);
- Boiling tests, etc.

Temporary Storage

This step should be as short as possible and better completely eliminated. Vegetables can be stored in:

- Simple stores, without artificial cooling;
- In refrigerated stores; or, in some cases,
- In silos (potatoes, etc.).

Simple stores should be covered, fairly cool, dry and well ventilated but without forced air circulation which can induce significant losses in weight through intensive water evaporation; air relative humidity should be at about 70-80per cent. Refrigerated storage is always preferable and in all cases a processing centre needs a cold room for this purpose, adapted in volume I capacity to the types and quantities of vegetables (and fruits) that are further processed.

Washing

Washing is used not only to remove field soil and surface micro-organisms but also to remove fungicides, insecticides and other pesticides, since there are laws specifying maximum levels of these materials that may be retained on the vegetable; and in most cases the allowable residual level is virtually zero. Washing water contains detergents or other sanitisers that can essentially completely remove these residues.

The washing equipment, like all equipment subsequently used, will depend upon the size, shape and fragility of the particular kind of vegetable:

- Flotation cleaner for peas and other small vegetables;
- Rotary washer in which vegetables are tumbled while they are sprayed with jets of water; this type of washer should not be used to clean fragile vegetables;

Sorting

This step covers two separate operations:

- Removal of non-standard vegetables (and fruit) and possible foreign bodies remaining after washing;
- Quality grading based on variety, dimensional, organoleptical and maturity stage criterion.

Skin Removal/Peeling

Some vegetables require skin removal. This can be done in various ways:

- Mechanical

  This type of operation is performed with various types of equipment which depend upon the result expected and the characteristics of the fruit and vegetables, for example:

  - A machine with abrasion device (potatoes, root vegetables);
  - Equipment with knives (apples, pears, potatoes, etc.);

- Equipment with rotating sieve drums (root vegetables). Sometimes this operation is simultaneous with washing (potatoes, carrots, etc.) or preceded by blanching (carrots).

- Chemical

  Skins can be softened from the underlying tissues by submerging vegetables in hot alkali solution. Lye may be used at a concentration of about 0.5-3per cent, at about 93° C (2000 F) for a short time period (0.5-3 min). The vegetables with loosened skins are then conveyed under high velocity jets of water which wash away the skins and residual lye. In order to avoid enzymatic browning, this chemical peeling is followed by a short boiling in water or an immersion in diluted citric acid solutions. It is more difficult to peel potatoes with this method because it is necessary to dissolve the cutin and this requires more concentrated lye solutions, up to 10per cent.

- Thermal

Wet Heat (Steam)

Other vegetables with thick skins such as beets, potatoes, carrots and sweet potatoes may be peeled with steam under pressure (about 10 at) as they pass through cylindrical vessels.

This softens the skin and the underlying tissue. When the pressure is suddenly released, steam under the skin expands and causes the skin to puff and crack. The skins are then washed away with jets of water at high pressure (up to 12 at).

Dry Heat (Flame)

Other vegetables such as onions and peppers are best skinned by exposing them to direct flame (about 1 min at 1000° C) or to hot gases in rotary tube flame peelers. Here too, heat causes steam to develop under skins and puff them so that they can be washed away with water.

Table. Losses at Vegetable Peeling, in per cent.

	Peeling methods		
Vegetables	Manual	Mechanical	Chemical
Potatoes	15-19	18-28	-
Carrots	13-15	16-18	8-10
Beets	1416	13-15	9-10

Manual peeling only use when the other methods are impossible or sometimes as a completion of the other three ways. Average losses at this step are given in Table.

Size Reduction

This step is applied according to specific vegetable and processing technology requirements.

Blanching

The special heat treatment to inactivate enzymes is known as blanching. Blanching is not indiscriminate heating. Too little is ineffective, and too much damages the vegetables by excessive cooking, especially where the fresh character of the vegetable is subsequently to be preserved by processing.

This heat treatment is applied according to and depends upon the specificity of vegetables, the objectives that are followed and the subsequent processing/ preservation methods.

Two of the more heat resistant enzymes important in vegetables are catalase and peroxidase. If these are destroyed then the other significant enzymes in vegetables also will have been inactivated. The heat treatment to destroy catalase and peroxidase in different vegetables are known, and sensitive chemical tests have been developed to detect the amounts of these enzymes that might survive a blanching treatment.

Because various types of vegetables differ in size, shape, heat conductivity, and the natural levels of their enzymes, blanching treatments have to be established on an experimental basis. As with sterilisation of foods in cans, the larger the food item the longer it takes for heat to reach the centre. Small vegetables may be adequately blanched in boiling water in a minute or two, large vegetables may require several minutes.

Blanching as a unit operation is a short time heating in water at temperatures of 100° C or below. Water blanching may be performed in double bottom kettles, in special baths with conveyor belts or in modern continuous blanching equipment.

In order to reduce losses of hydrosoluble substances (mineral salts, vitamins, sugars, etc.) occurring during water blanching, several methods have been developed:

- Temperature setting at 85-95° C instead of 100° C;
- Blanching time has to be just sufficient to inactivate enzymes catalase and peroxidase;
- Assure elimination of air from tissues.

An illustration of blanching parametres is seen in Table given below.

Table. Blanching Parametres for Some Vegetables.

Vegetables	Temperature, °C	Time, min.
Peas	85-90	2-7
Green beans	90-95	2-5
Cauliflower	Boiling	2
Carrots	90	3-5
Peppers	90	3

Steam heat treatment can also be applied instead of water blanching as a preliminary step before freezing or drying, as long as the preservation method is only used for enzyme inactivation and not to modify consistency. For drying, the vegetables are conveyed directly from steaming equipment to drying installations without cooling. Vegetable steaming is carried out in continuous installations with conveyer belts made from metallic sieves.

Cooling of vegetables after water blanching or steaming is performed in order to avoid excessive softening of the tissues and has to follow immediately after these operations; one exception is the case of vegetables for drying which can be transferred directly to drying equipment without cooling.

Natural cooling is not recommended because is too long and generates significant losses in vitamin C content. Cooling in pre-cooled air (from special installations) is sometimes used for vegetables that will be frozen

Cooling in water can be achieved by sprays or by immersion; in any case the vegetables have to reach a temperature value under 37° C as soon as possible. Too long a cooling time generates supplementary losses in valuable hydrosoluble substances; in order to avoid this, the temperature of the cooling water has to be as low as possible.

Canning

Large quantities of vegetable products are canned. A typical flow sheet for a vegetable canning operation (which also applies to fruit for the most part) covers some food process unit operations performed in sequence: harvesting; receiving; washing; grading; heat blanching; peeling and coring; can filling; removal of air under vacuum; sealing/closing, retorting/heat treatment; cooling; labelling and packing. The vegetable may be canned whole, diced, purėed, as juice and so on.

On-line Simplified Methods for Enzyme Activity Check

Peroxidase test:

- Solutions. In order to check the peroxidase activity two solutions have to be prepared:
  - 1per cent guaiacol in alcohol solution (1 g guaiacol is dissolved in about 50 cm^3 of 96per cent ethylic alcohol and then this preparation is brought to 100 cm° with the same solvent);
  - Peroxide solution 0.3per cent (1 cm^3 perhydrol is brought to 100 cm^3 with distilled water.
- Sampling. From various parts of the material samples are taken (about 20-30 pieces, etc.); the material is then crushed in a labouratory bowl in order to obtain an average sample.
- Check. Prom the average sample, 10-20 g of material is taken in a medium capacity test tube; on this sample are poured: 20 cm^3 distilled water; 1 cm^3 of 1per cent guaiacol solution; 1.6 cm^3 of peroxide solution.

The contents of the test tube is shaken well. The gradual appearance of a weak pink colour indicates an incomplete peroxidase inactivation - reaction slightly positive. If there are no tissue colour modifications after 5 minutes, the reaction is negative and the enzymes have been inactivated.

As an orientative check it is also possible to simply pour a few drops of 1per cent guaiacol solution and 0.3per cent peroxide solution directly on blanched and crushed vegetables. A rapid and intensive brown-reddish tissue colouring indicates a high peroxidase activity (positive reaction).

Catalase Test

In order to identify the catalase enzyme activity, 2 g of dehydrated vegetables are well crushed and mixed with about 20 cm^3 of distilled water. After 15 min softening, 0.5 cm^3 of a 0.5per cent or 1per cent peroxide solution is poured on prepared vegetables. In the presence of catalase, a strong oxygen generation is observed for about 2-3 minutes.

These tests are of a paramount importance in order to determine the vegetable blanching treatments (temperature and time); incomplete enzyme inactivation has a negative effect on finished product quality.

For cabbage catalase inactivation by blanching is sufficient; blanching further to peroxidase inactivation would have negative effects on product quality and even complete browning. For all other vegetables and for potatoes, both tests MUST be negative, for catalase and for peroxidase.

Fresh Vegetable Storage

The vegetables can be stored, in some specific natural conditions, in fresh state, that is without significant modifications of their initial organoleptic properties. Fresh vegetable storage can be short term; this was briefly covered under temporary storage before processing. Also fresh vegetable storage can be long term during the cold season in some countries and in this case it is an important method for vegetable preservation in the natural state. In order to assure preservation in long term storage, it is necessary to reduce respiration and transpiration intensity to a minimum possible; this can be achieved by:

- Maintenance of as low a temperature as possible (down to 0° C),
- Air relative humidity increased up to 85-95 per cent and
- CO_2 percentage in air related to the vegetable species.

Vegetables for storage must conform to following conditions: they must be of one of the autumn or winter type variety; be at edible maturity without going past this stage; be harvested during dry days; be protected from rain, sun heat or wind; be in a sound state and clean from soil; be undamaged.

From the time of harvest and during all the period of their storage vegetables are subject to respiration and transpiration and this is on account of their reserve substances and water content. The more the intensity of these two natural processes are reduced, the longer sound storage time will be and the more losses will be reduced. For this reason, vegetables have to be handled and transported as soon as possible in the storage conditions (optimal temperature and air relative humidity for the given species). Even in these optimal conditions storage will generate losses in weight which are variable and depend upon the species.

VEGETABLE NATURAL ACIDIFICATION TECHNOLOGY

Raw materials must follow strict specifications for a high quality finished product; the following parametres must be considered as critical:

- Adapt a uniform size according to the finished product requirements; for example, gherkins will need to have a maximum length of 9 cm for raw vegetables. Generally 15 cm size/length will be a maximum for high quality cucumber products in many countries. However, according to local preferences, bigger cucumbers could be also in demand.
- Cylindrical or ovoidal shape;
- Dark green colour;
- Absence of surface defects due to cryptogamic diseases.

Cucumbers have to be picked at their ripeness for eating, when the sugar content is at about 1.5-2.2per cent, needed for lactic fermentation. Unripe cucumber does not have enough sugar.

- Small holes are made in large size cucumbers skin;
- *Rreceptacle filling:* raw material is simply put in the receptacles in bulk, with care to arrange them in such a way that a maximum of pieces could be introduced;
- *Salt solution preparation:* 6per cent salt solution (NaCl);
- *Salt solution addition*: the salt solution is poured into the receptacle;
- Fermentation is carried out at 20-30° C, anaerobically. This step takes generally 4-8 weeks. Acidity reaches a value up to 1.5per cent lactic acid (and in some exceptional cases up to 2per cent lactic acid) which corresponds to a maximum pH value of 4.1.
- Storage; after the last fermentation stage, drums and other receptacles have to be stored at low temperature; best conditions for 12 months shelf life should be below + 15° C. Storage temperature will determine the shelf life of the products.
- Addition of 1000 ppm potassium sorbate will prevent mould development without having any influence on lactic fermentation.
- Raw material grading by size is a very important technological step. In order to accelerate brine penetration, mainly for medium to large size cucumbers, the practice of making small holes in the raw material skins is generally recommended.

- A major factor influencing the quality of lactic fermented cucumbers is the water durity; optimal results are obtained at 15-20° durity.
- Cucumber consistency/texture is influenced by the formation of calcium pectate with the pectic substances from raw material tissues. In some countries, calcium chloride (0.3-0.5 per cent) is added in order to firm up the cucumber consistency. Chlorinated water which still contains active chlorine can inhibit or even stop the lactic fermentation.

Sauerkraut

In some countries cabbages are submitted to lactic fermentation as whole vegetables; however, in many countries the cabbage is shredded before fermentation.

As shredded cabbage and its technology is at the basis of an important industry, giving good quality products, with a uniform fermented product and with good keeping quality and ease of distribution, this will be described first.

Cabbage as raw material for sauerkraut must be sound, ripe for eating, well-leafed and from suitable varieties.

Optimum total sugar level needed for the lactic fermentation is 24per cent; generally good quality raw material contains up to 30-60 mg/100 g of vitamin C.

Removal of External Leaves

Coring is done with a specially adapted mechanical screw; this operation generates small particles of finely divided cabbage which will be mixed with the main part of vegetable during shredding/chopping.

The core represents about 10per cent from the whole cabbage, is rich in sugar and vitamin C, but being too high in fibre content needs to be chopped separately as described.

Shredding/Cutting of cabbage is carried out with complex specific equipment which is generally installed directly on the "top" of fermentation silos and is mobile, installed on rails and moves all along the silos. The dimension of resulting shredded cabbage is about 2-3 mm thick.

The same complex equipment is designed to grind the added salt to fine particles and to distribute shredded cabbage and ground salt in an uniform manner to the fermentation silos.

The usual capacity of fermentation silos is up to 30 tons, with separate compartments of 45 tons each.

Salt addition is carried out by the equipment described above; the proportion of salt is 2-2.5per cent with respect to cabbage.

This proportion must not be changed because the salt in this technology does not have a preservative role but only that to extract from cabbage the juice needed for fermentation. It is be preferable to obtain a fairly light pressure on cabbage just after salt addition with some simple mechanical means.

This is important in order to:

- Create an anaerobic medium for fermentation;
- Facilitate external diffusion of cellular juice;
- Assure a rational use of the fermentation space.

Fermentation: The maximum acidity level obtained is generally of about 1.5per cent lactic acid (and very rarely 2.5per cent); this is obtained in 4-6 weeks. Optimal acidity is 1.0-1.8per cent and pH value 4.1 or lower.

Fermentation temperature is at 20-25° C in the first phase and needs to be lowered then to 14-18° C.

During fermentation, the brine from each storage/fermentation silo cell is periodically circulated with a pump in order to uniformise the fermentation process.

Storage is performed in same silos used for fermentation, or the finished products is removed from silos and packed in drums and other receptacles according to distribution schedule.

These silos are usually made of reinforced concrete and coated with gritstone plates or with an acid-resisting material layer. At small scale and in traditional processing, shredded sauerkraut can be obtained by using simple available glass or rigid plastic receptacles.

At home, this process can use glass jars and/or local/traditional pottery receptacles from a minimum size of 2-3 kg up to the available/practical sizes (better limited to 10-15 kg).

In some countries shredded sauerkraut is preserved in receptacles by pasteurization, once the fermentation process has been completed.

Whole Sauerkraut

According to the consumer preference in different countries and to the specific situations it is also usual to preserve whole cabbages by lactic fermentation. At small or medium scale operations, whole cabbage could be processed/fermented in cylindric receptacles like 30 to 200 litre rigid plastic drums, or rectangular receptacles made from food grade rigid plastic.

It is possible to find this type of drum in a significant number of developing countries. These two types of rigid plastic receptacles could also be used for shredded sauerkraut production.

Prepared whole cabbages are put into fermentation receptacles and a 5-6 per cent salt concentration brine is poured on top.

The fermentation conditions are the same as for shredded sauerkraut. In order to assure a uniform fermentation and to avoid a strict anaerobic (butyric) fermentation it is necessary to apply a periodic juice "aeration" (each 2-3 days at the beginning of the fermentation, and then each 5-7 days).

Other Acidified Vegetables

In principle all vegetables with a sugar content of at least 2 per cent could be preserved by lactic fermentation. From a practical point of view it is mainly the following vegetables which are preserved by this technology: unripe tomatoes (green tomatoes), peppers, eggplant, carrots and cauliflower, alone or usually in a mix with cucumber as mixed pickles.

Fermentation of individual vegetables is carried out according to a flow-sheet as described for whole sauerkraut. The type of cut, brine concentration and frequency of operating steps have to be adapted to each case; green tomatoes are fermented as whole vegetable.

9

Preservation of Fruits and Vegetables

FRUIT PRESERVES

Fruit preserves are fruits, or vegetables, that have been prepared and canned or sealed air-tight for long term storage. The preparation of fruit preserves traditionally involves the use of pectin as a gelling agent, although sugar or honey may be used as well. The ingredients used and how they are prepared will determine the type of preserves; jams, jellies and marmalades are all examples of different styles of fruit preserves that vary based upon the ingredients used. There are various varieties of fruit preserves made globally, and they can be made from sweet or savory ingredients. In North America, the plural form preserves is used, while the singular preserve is used in British and Commonwealth English. Additionally, the name of the type of fruit preserves will also vary depending on the regional variant of English being used.

VARIATIONS

Confit

Confit, which is the past participle form of the French verb "confire" or "to preserve", is most often applied to preservation of meats, especially poultry and pork, by cooking them in their own fat or oils and allowing the fats to set. However, the term can also refer to fruit or vegetables which have been seasoned and cooked with honey or sugar until it has reached a jam-like consistency. Savory confits, such as ones made with garlic or tomatoes, may call for a savory oil such as virgin olive oil as the preserving agent.

Conserves

A conserve is a jam made of fruit stewed in sugar. Often the making of conserves can be trickier than making a standard jam, because the balance between cooking, or sometimes steeping in the hot sugar mixture for just enough time to allow the flavour to be extracted from the fruit, and sugar to penetrate the fruit, and cooking too long that fruit will break down and liquefy.

This process can also be achieved by spreading the dry sugar over raw fruit in layers, and leaving for several hours to steep into the fruit then just heating the resulting mixture only to bring to the setting point. As a result of this minimal cooking, some fruits are not particularly suitable for making into conserves, because they require cooking for longer periods to avoid issues such as tough skins. Currants and gooseberries, and a number of plums are among these fruits. Due to this shorter cooking period, not as much pectin will be released from the fruit, and as such, conserves will sometimes be slightly softer set than some jams. An alternate definition holds that conserves are preserves made from a mixture of fruits and/or vegetables. Conserves may also include dried fruit or nuts.

Fruit Butter

Fruit butter, in this context, refers to a process where the whole fruit is forced through a sieve or blended after the heating process:

- "Fruit butters are generally made from larger fruits, such as apples, plums, peaches or grapes. Cook until softened and run through a sieve to give a smooth consistency. After sieving, cook the pulp... add sugar and cook as rapidly as possible with constant stirring.... The finished product should mound up when dropped from a spoon, but should not cut like jelly. Neither should there be any free liquid."

Fruit Curd

Fruit curd is a dessert topping and spread usually made with lemon, lime, orange, or raspberry. The basic ingredients are beaten egg yolks, sugar, fruit juice and zest which are gently cooked together until thick and then allowed to cool, forming a soft, smooth, intensely flavoured spread. Some recipes also include egg whites and/or butter.

Fruit Spread

Fruit spread refers to a jam or preserve with no added sugar.

Jam

Jam contains both fruit juice and pieces of the fruit's flesh, however, some cookbooks define jam as cooked and gelled fruit purees. Properly, the term jam refers to a product made with whole fruit, cut into pieces or crushed. The fruit is heated with water and sugar to activate the pectin in the fruit. The mixture is then put into containers. The following extract from a US cookbook describes the process:

- "Jams are usually made from pulp and juice of one fruit, rather than a combination of several fruits. Berries and other small fruits are most frequently used, though larger fruits such as apricots, peaches, or plums cut into small pieces or crushed are also used for jams.

Good jam has a soft even consistency without distinct pieces of fruit, a bright colour, a good fruit flavour and a semi-jellied texture that is easy to spread but has no free liquid."

Jelly

Jelly is a clear or translucent fruit spread made from sweetened fruit juice and set using naturally occurring pectin. Additional pectin may be added where the original fruit does not supply enough, for example with grapes. Jelly can be made from sweet, savory or hot ingredients. It is made by a process similar to that used for making jam, with the additional step of filtering out the fruit pulp after the initial heating. A muslin or stockinette "jelly bag" is traditionally used as a filter, suspended by string over a bowl to allow the straining to occur gently under gravity. It is important not to attempt to force the straining process, for example by squeezing the mass of fruit in the muslin, or the clarity of the resulting jelly will be compromised.

- "Good jelly is clear and sparkling and has a fresh flavour of the fruit from which it is made. It is tender enough to quiver when moved, but holds angles when cut.
- *Extracting Juice*: Pectin is best extracted from the fruit by heat, therefore cook the fruit until soft before straining to obtain the juice.... Pour cooked fruit into a jelly bag which has been wrung out of cold water. Hang up and let drain. When dripping has ceased the bag may be squeezed to remove remaining juice, but this may cause cloudy jelly."

Marmalade

British-style marmalade is a sweet preserve with a bitter tang made from fruit, sugar, water, and a gelling agent. American-style marmalade is sweet, not bitter. In English-speaking usage, "marmalade" almost always refers to a preserve derived from a citrus fruit, most commonly oranges, although onion marmalade is also used as an accompaniment to savoury dishes. The recipe includes sliced or chopped fruit peel, which is simmered in fruit juice and water until soft; indeed marmalade is sometimes described as jam with fruit peel. Such marmalade is most often consumed on toasted bread for breakfast.

The favoured citrus fruit for marmalade production in the UK is the "Seville orange", *Citrus aurantium* var. aurantium, thus called because it was originally imported from Seville in Spain; it is higher in pectin than sweet oranges, and therefore gives a good set. Marmalade can also be made from lemons, limes, grapefruit, strawberries or a combination.

REGIONAL TERMINOLOGY

The term preserves is usually interchangeable with jam. Some cookbooks define preserves as cooked and gelled whole fruit which includes a significant portion of the fruit. The terms jam and jelly are used in different parts of the English-speaking world in different ways. In the United States, both jam and jelly are sometimes popularly referred to as "jelly", whereas in the United Kingdom, Canada, India and Australia, the two terms are more strictly differentiated. In Australia and South Africa, the term jam is more popularly used as a generic term for both jam and jelly. To further confuse the issue, the term jelly is also used in the UK, South Africa, Australia, India and New Zealand to refer to a gelatin dessert, known in North America as jello, derived from the brand name Jell-O.

PRODUCTION

In general, jam is produced by taking mashed or chopped fruit or vegetable pulp and boiling it with sugar and water. The proportion of sugar and fruit varies just as to the type of fruit and its ripeness, but a rough starting point is equal weights of each. When the mixture reaches a temperature of 104°C, the acid and the pectin in the fruit react with the sugar, and the jam will set on cooling. However, most cooks work by trial and error, bringing the mixture to a "fast rolling boil", watching to see if the seething mass changes texture, and dropping small samples on a plate to see if they run or set. Commercially produced jams are usually produced using one of two methods. The first is the open pan method, which is essentially a larger scale version of the method a home jam maker would use. This gives a traditional flavour, with some caramelization of the sugars.

The second commercial process involves the use of a vacuum vessel, where the jam is placed under a vacuum, which has the effect of reducing its boiling temperature to anywhere between 65-80°C depending on the recipe and the end result desired. The lower boiling temperature enables the water to be driven off as it would be when using the traditional open pan method, but with the added benefit of retaining more of the volatile flavour compounds from the fruit, preventing caramelization of the sugars, and of course reducing the overall energy required to make the product. However, once the desired amount of water has been driven off, the jam still needs to be heated briefly to 95-100 °C to kill off any micro-organisms that may be present; the vacuum pan method does not kill them all.

During the commercial filling of the jam into jars, it is common to use a flame to sterilize the rim of the jar and the lid to destroy any yeasts and molds which may cause spoilage during storage. It is also common practice to inject steam into the head space at the top of the jar immediately prior to the fitting of the lid, in order to create a vacuum. Not only does this vacuum help prevent the growth of spoilage organisms, it also pulls down the tamper evident safety button when lids of this type are employed. How easily a jam sets depends on the pectin content of the fruit.

Some fruits, such as gooseberries, redcurrants, blackcurrants, most citrus fruits, apples, and raspberries, set very well; others, such as strawberries and ripe blackberries, often need to have pectin added. There are commercial pectin products on the market, and most industrially-produced jams use them. Home jam-makers sometimes rely on adding a pectin-rich fruit to a poor setter, for example apple to blackberries. Other tricks include extracting juice from lemons, redcurrants or gooseberries, or making a pectin stock with whole apples or just the cores and skins; once cooled, this "stock" can then be frozen for later use. Making jam at home is a popular handicraft activity, and many take part in this. Homemade jam may be made for personal consumption, or as part of a cottage industry.

ASPIC

Aspic is a dish in which ingredients are set into a gelatin made from a meat stock or consommé. Similar dishes, made with commercial gelatin mixes instead of stock or consommé, are usually called gelatin salads. When cooled, stock that is made from meat congeals because of the natural gelatin found in the meat. The stock can be clarified with egg whites, and then filled and flavoured just before the aspic sets. Almost any type of food can be set into aspics.

Most common are meat pieces, fruits, or vegetables. Aspics are usually served on cold plates so that the gel will not melt before being eaten. A meat jelly that includes cream is called a chaud-froid. Nearly any type of meat can be used to make the gelatin: pork, beef, veal, chicken, turkey, or fish. Gelatin is also found in cartilage. The aspic may need additional gelatin in order to set properly. Veal stock provides a great deal of gelatin; in making stock, veal is often included with other meat for that reason. Fish consommés usually have too little natural gelatin, so the fish stock may be double-cooked or supplemented. Since fish gelatin melts at a lower temperature than gelatins of other meats, fish aspic is more delicate and melts more readily in the mouth. Vegetables and fish stocks need gelatin to create a mould.

Historically, meat jellies were made before fruit and vegetable jellies. By the Middle Ages at the latest, cooks had discovered that a thickened meat broth could be made into a jelly. A detailed recipe for aspic is found in Le Viandier, written in around 1375. In the 18th century, Marie-Antoine Carême created chaud froid in France. Chaud froid means "hot cold" in French, referring to foods that were prepared hot and served cold. Aspic was used an chaud froid sauce in many cold fish and poultry meals. The sauce added moisture and flavour to the food. Marie-Antoine Carême also invented various types of aspic and ways of preparing aspic.

Aspic didn't come into prominence until the early 20th century. Aspic was used to hold meats and prevented them from becoming spoiled. The gelatin prevented air and bacteria from ruining cooked meat. It wasn't eaten until the 1950s, when it became popular dish. In the 1950s, aspic was a popular dinner staple throughout the United States. Jello was popular in the 1950s and the demands for meat were abundant. One of the most popular dishes were the tomato aspic. Cooks used to show off aesthetic skills by creating inventive aspics.

USES

Aspic can also be referred as a aspic gelèe or aspic jelly. Aspic jelly may be colourless or contain various shades of amber. Aspic can be used to protect food from the air, give food more flavour, or as a decoration. There are three types of aspic textures: delicate, sliceable, and inedible.

The delicate aspic is soft. The sliceable aspic must be made in a terrine or in a aspic mould. It is more firmer than the delicate aspic. The inedible aspic is never for consumption. It is usually for decoration. Aspic is often used to glaze food pieces in food competitions to make the food glisten and making it more appealing to the eye. Foods dipped in aspic can have a lacquered finish for a fancy presentation. Aspic can be cut into various shapes and be used as a garnish for deli meats or pâtés.

JUGGING

In cooking, jugging is the process of stewing meat for a long time in a tightly covered container, such as a casserole or an earthenware jug. Sometimes the cooking liquid includes some of the animal's blood. In French, such a stew of a game animal, thickened with the animal's blood, is known as a civet. One common traditional dish that involves jugging is Jugged Hare which is a whole hare, cut into pieces, marinated and cooked with red wine and juniper berries in a tall jug that stands in a pan of water. It is traditionally served with the hare's blood and Port wine.

The recipe goes on to describe cooking the pieces of hare in water in a jug that is set within a bath of boiling water to cook for three hours. Beginning in the nineteenth century, Glasse has been widely credited with having started the recipe with the words "First, catch your hare". This attribution is apocryphal. However, having a freshly caught, or shot, hare enables one to obtain its blood.

A freshly killed hare is prepared for jugging by removing its entrails and then hanging it in a larder by its hind legs, which causes the blood to accumulate in the chest cavity. One method of preserving the blood after draining it from the hare is to mix it with red wine vinegar in order to prevent it coagulating, and then to store it in a freezer.

Many other British cookbooks from before the middle of the 20th century have recipes for Jugged Hare. Merle and Reitch have this to say about Jugged Hare, for example:

- The best part of the hare, when roasted, is the loin and the thick part of the hind leg; the other parts are only fit for stewing, hashing, or jugging. It is usual to roast a hare first, and to stew or jug the portion which is not eaten the first day. To Jug a Hare: This mode of cooking a hare is very desirable when there is any doubt as to its age, as an old hare, which would be otherwise uneatable, may be made into an agreeable dish. In 2006, a survey of 2021 people for the television channel UKTV Food found that only 1.6 per cent of the people aged under 25 recognized Jugged Hare by name. 7 out of 10 of those people stated that they would refuse to eat Jugged Hare if it were served at the house of a friend or a relative. Jugged Rabbit is an alternative to Jugged Hare. It is considered a speciality of the cuisine of Martinique. Another jugged dish, also traditional in the United Kingdom, is Jugged Kippers, which is kippers in a covered jug, cooked in boiling water.

MODIFIED ATMOSPHERE

Modified atmosphere is an uncommon technical definition that describes the practice of modifying the composition of the internal atmosphere of a package in order to improve the shelf life.

The modification process often tries to lower the amount of oxygen, moving it from 20 per cent to 0 per cent, In order to slow down the growth of aerobic organisms and the speed of oxidation reactions.

The removed oxygen can be replaced with nitrogen, commonly acknowledged as an inert gas, or carbon dioxide, which can lower the pH or inhibit the growth of bacteria. Carbon monoxide can be used for keeping the red colour of meat.

Re-balancing of gases inside the packaging can be achieved using active techniques such as gas flushing and compensated vacuum or passively by designing "breathable" films known as equilibrium modified atmosphere packaging.

SCIENTIFIC TERMS

- MAP = Modified atmosphere packaging
- EMAP = Equilibrium modified atmosphere packaging
- MA/MH = Modified atmosphere/modified humidity packaging.

PRODUCTS

Under MA products like red meat, seafood, minimally processed fruits and vegetables, pasta, cheese, bakery goods, poultry, cooked and cured meats, ready meals and dried foods are packaged. The three major commodity types are fruits and vegetables, meat and meat products and seafood. It has been estimated that 25-40 per cent of all fresh produce harvested will not reach the consumers table, due to spoilage and mishandling that occurs during distribution.

MODIFIED ATMOSPHERE PACKAGING

Modified Atmosphere Packaging is a technique used for prolonging the shelf-life period of fresh or minimally processed foods. In this preservation technique the air surrounding the food in the package is changed to another composition.

This way the initial fresh state of the product may be prolonged. It is the shelf-life of perishable products like meat, fish, fruits and vegetables that will be prolonged with MAP since it slows the natural deterioration of the product. MAP is used with various types of products, where the mixture of gases In the package depends on the type of product, packaging materials and storage temperature.

Meat and fish need very low gas permeability films so for non-respiring products high barrier films are used. The initial flushed gas-mixture will be maintained inside the MA package. But fruits and vegetables are respiring products where the interaction of the packaging material with the product is important. If the permeability of the packaging film is adapted to the products respiration, an equilibrium modified atmosphere will establish in the package and the shelf-life of the product will increase.

EQUILIBRIUM MODIFIED ATMOSPHERE PACKAGING

Among fresh-cut produce Equilibrium Modified Atmosphere Packaging is the most commonly used packaging technology. When packaging vegetables and fruits the gas atmosphere of package is not air but consists usually of a lowered level of O_2 and a heightened level of CO_2.

This kind of package slows down the normal respiration of the product and so to prolong the shelf-life of the product. Of course there are other factors like the size of the product, severity of preparation, maturity of the product and type of tissue that have an effect to the shelf-life of an EMA packaged produce.

TECHNOLOGY

There are two techniques used in the industry to pack vegetables namely gas-flushing and compensated vacuum. For its cheapness the gas-flushing is more widely used. In gas-flushing the package is flushed with a desired gas mixture, as in compensated vacuum the air is removed totally and the desired gas mixture then inserted.

The label "packaged in a protective atmosphere" can refer to either of these; an example of a gas mixture used for non-vegetable packaged food is

99.9 per cent nitrogen gas, which is inert at the temperatures and pressures the packaging is subjected to.

GASES

The atmosphere in an MA package consists of N_2, O_2, CO_2. It is the altered ratio of these gases that makes a difference in the prolongation of shelf life. By reducing the O_2-level and increasing the CO_2-level, ripening of fruits and vegetables can be delayed, respiration and ethylene production rates can be reduced, softening can be retarded and various compositional changes associated with ripening can be slowed down. Oxygen is essential when packaging fresh fruits and vegetables as they continue to respire after harvesting.

The absence of O_2 can lead to anaerobic respiration in the package which accelerates senescence and spoilage. Too high levels of O_2 do not retard respiration significantly and it is around 12 per cent of O_2 where the respiration rate starts to decrease. So oxygen is used in low levels for positive effect. When packaging meat and fish, the high CO_2-levels are effective bacterial and fungal growth inhibitors.

In the case of vegetables and fruits, CO_2 is not a major factor since CO_2-levels above 10 per cent are needed to suppress fungal growth significantly. Unfortunately higher levels than 10 per cent of CO_2 are working phytotoxic for fresh produce. Nitrogen is used as a filler gas since it neither encourages or discourages bacterial growth.

PACKAGING FILMS

When selecting packaging films for EMAP of fruits and vegetables the main characteristics to consider are gas permeability, water vapour transmission rate, mechanical properties, transparency, type of package and sealing reliability. Traditionally used packaging films like LDPE, PVC, EVA and OPP are not permeable enough for highly respiring products like fresh-cut produces, mushrooms and broccoli. As fruits and vegetables are respiring products, there is a need to transmit gases from and to the package. Films designed with these properties are called permeable films. Other films, called barrier films, are designed to prevent the exchange of gases and are mainly used with non-respiring products like meat and fish. EMAP films developed to control the humidity level as well as the gas composition in the sealed package are beneficial for the prolonged storage of fresh fruits, vegetables and herbs that are sensitive to moisture.

PRESERVATION OF VEGETABLES BY ACIDIFICATION

Food acidification is a means of preventing their deterioration in so far as a non-favourable medium for micro-organisms development is created. This acidification can be obtained by two ways: natural acidification and artificial acidification.

Natural Acidification

This is achieved by a predominant lactic fermentation which assures the preservation based on acidoceno-anabiosys principle; preservation by lactic fermentation is called also biochemical preservation. Throughout recorded history food has been preserved by fermentation. In spite of the introduction of modern preservation methods, lactic acid fermented vegetables still enjoy a great popularity, mainly because of their nutritional and gastronomic qualities.

The various preservation methods discussed thus far, based on the application of heat, removal of water, cold and other principles, all have the common objective of decreasing the number of living organisms in foods or at least holding them in check against further multiplication.

Fermentation processes for preservation purposes, in contrast, encourage the multiplication of micro-organisms and their metabolic activities in foods. But the organisms that are encouraged are from a select group and their metabolic activities and end products are highly desirable. The extent of this desirability is emphasised by a partial list of fermented fruits and vegetable products from various parts of the world.

There are some characteristic features in the production of fermented vegetables which will be pointed out below using cucumbers as an example. In the production of lactic acid fermented cucumbers, the raw material is put into a brine without previous heating. Through the effect of salt and oxygen deficiency the cucumber tissues gradually die. At the same time, the semi-permeability of the cell membranes is lost, whereby soluble cell components diffuse into the brine and serve as food substrate for the micro-organisms.

Under such specific conditions of the brine the lactic acid bacteria succeed in overcoming the accompanying micro-organisms and lactic acid as the main metabolic products is formed. Under favourable conditions (for example moderate salt in the brine, use of starter cultures) it takes at least 3 days until the critical pH value of 4.1 or less - desired for microbiological reasons - is reached.

Beside the typical taste, for the consumer a crisp texture is the most important quality criterion for fermented vegetables. Because there is no heating step before the fermentation, the indigenous plant enzymes in the fermenting materials are still present during the very first phase. After the destruction of the cell membranes they easily get to their active sites and under favourable conditions they can easily cause softening.

The environmental conditions act in a different manner on single enzymes or enzymes systems: some enzymes are strongly inhibited by salt, others are activated, and in the acid pH-region many enzymes are irreversibly inactivated. Beside indigenous enzymes also enzymes produced by micro-organisms can be responsible for the undesired soft products. In technically advanced societies the major importance of fermented foods has come to be variety they add to the diet. However, in many less developed areas of the world, fermentation and natural drying are the major food preservation methods and as such are vital to survival of a large proportion of the world's current population.

Artificial acidification is carried out by adding acetic acid which is the only organic acid harmless for human health and stable in specific working conditions; in this case biological principles of the preservation are acidoanabiosys and, to a lesser extent, acidoabiosys.

Combined acidification is a preservation technology which involves as a preliminary processing step a weak lactic fermentation followed by acidification (vinegar addition). The two main classes of vegetables preserved by acidification are sauerkraut and pickles; the definitions of these products adapted from US Code of Federal Register are as follows.

Bulk sauerkraut. Bulk or barrelled sauerkraut is the product of characteristic acid flavour, obtained by the full fermentation, chiefly lactic, of properly prepared and shredded cabbage in the presence of 2-3per cent salt. On completion of fermentation, it contains not more than 1.5per cent of acid, expressed as lactic acid.

Canned sauerkraut. Canned (or packaged) sauerkraut, is prepared from clean, sound, well-matured heads of the cabbage plant (Brassica oleracea var. capitata L.) which have been properly trimmed and cut; to which salt is added and which is cured by natural fermentation. The product may or may not be packed with pickled peppers, pimientos, or tomatoes or contain other flavouring ingredients to give the product specific flavour characteristics. The product:

- May be canned by processing sufficiently by heat to assure preservation in hermetically sealed containers; or
- May be packaged in sealed containers and preserved with or without the addition of benzoate of soda or any other ingredient permissible under the provisions of Food and Drug Administration (FDA).

Pickles. "Pickles" means the product prepared entirely or predominantly from cucumbers (Cucumis sativus L.). Clean, sound ingredients are used which may or may not have been previously subjected to fermentation and curing in a salt brine (solution of sodium chloride, NaCl).

The prepared pickles are packed in a vinegar solution to which may be added salt and other vegetables, nutritive sweeteners, seasonings, flavourings, spices, and other ingredients permissible under FDA regulations. The product is packed in suitable containers and heat treated, or otherwise processed to assure preservation. Sauerkraut and pickle products can be preserved under the effect of natural or added acidity, followed by pasteurization when this acidification is not sufficient. Sauerkraut is a very good source of vitamin C; the importance of this product should be emphasised in developing countries as a simple technology which can be applied mainly for consumption of the finished products in remote, isolated areas during the cold season. It is also a excellent technology to be learned to schools which have their own source of cabbage and cucumbers through school agricultural farms.

Sauerkraut and pickles are manufactured on an industrial scale in significant quantities world-wide. However, the basic technology is simple and could be applied at home, farm and community level after some explanation and training. The natural acidification preservation could be considered similar to sun/solar drying in terms of training and development.

Table. Some Industrial Fermentation Processes in Food Industries.

I. Lactic acid bacteria	
— cucumbers	dill pickles, sour pickles
— cabbage	sauerkraut
— turnips	sauerruben
— lettuce	lettuce kraut
— mixed vegetables, turnips, radish, cabbage	
— mixed Chinese vegetables, cabbage	Kimchi
— vegetables and milk	Tarhana
— vegetables and rice	Sajur asin
II. Lactic acid bacteria with other micro-organisms	
— with yeasts	Nukamiso pickles
— with moulds	tempeh, soy sauce
III.Acetic acid bacteria - wine, cider or any alcoholic and sugary or starchy products may be converted to vinegar	
IV.Yeasts	
— fruit	wine, vermouth

The principle of this technology is to add sugar in a quantity that is necessary to augment the osmotic pressure of the product's liquid phase at a level which will prevent microorganism development. From a practical point of view, however, it is usual to partially remove water (by boiling) from the product to be preserved, with the objective of obtaining a higher sugar concentration. In concentrations of 60per cent in the finished products, the sugar generally assures food preservation. It is important to know the ratio between the total sugar quantity in the finished product and the total sugar concentration in the liquid phase because this determines, in practice, the sugar preserving action. The percent composition of a product preserved with sugar, for example marmalade, can be expressed as follows: [I + S + S + N + W] = 100;

I = insoluble substance;

S = sugar from fruits;

S= added sucrose;

N= soluble "non sugar"

W= water.

In this case, total sugar concentration, in the liquid phase, of the finished product is:

$$X = 100 (S + s)/100 - (n + i) \text{ [per cent]}$$

Therefore, in the case of a standard marmalade with 55 per cent sugar added (calculated on the finished product basis), the real concentration in the liquid phase is for example:

$$X = 100 (55 + 8)/100 - (5 + 3) = 68.5 \text{per cent}$$

In the food preservation with sugar, the water activity cannot be reduced below 0.845; this value is sufficient for bacteria and neosmophile yeast inhibition but does not prevent mould attack. For this reason, various means are used to avoid mould development:

- Finished product pasteurization (jams, jellies, etc.);
- Use of chemical preservatives in order to obtain the antiseptisation of the product surface.

It is very important from a practical point of view to avoid any product contamination after boiling and to assure an hygienic operation of the whole technological process (this will contribute to the prevention of product moulding or fermentation).

Storage of the finished products in good conditions can only be achieved by ensuring the above level of water activity.

METHODS OF REDUCING DETERIORATION

A knowledge of deterioration factors and the way they act, including the rates of deterioration to a specific category of food, means that it is possible to list the ways of lowering or stopping the action and obtaining fruit and vegetable preservation.

In order to maintain their nutritional value and organoleptic properties and because of technical-economical considerations, not all the identified means against deterioration actually have practical applications for fruit and vegetable preservation.

Technical methods of reducing food deterioration

This classification of methods of reducing deterioration presents some difficulties because their preservation effects are physical, physico-chemical, chemical and biochemical complex phenomena which rarely act in isolation. Normally they take place together or one after the other.

From the whole list of possible methods of reducing deterioration, over the years, some procedures for fruit and vegetable preservation have found practical application. These technical means can be summarised as follows:

Physical	Heating
Cooling	Lowering of water content Drying/dehydration.
Concentration	Sterilising filtration
Irradiation	Other physical means (high pressure, vacuum, inert gases)
Chemical	Salting
Smoking	Sugar additio
Artificial acidification	Ethyl alcohol addition
	Antiseptic substance action
Biochemical	Lactic fermentation (natural acidification)
	Alcoholic fermentation

Table. Procedures for Fruit and Vegetable Preservation.

Procedures	Practical applications
Fresh storage	Fruits, vegetables
Cold storage	Fruits, vegetables
Freezing	Fruits, vegetables
Drying/dehydration	Fruits, vegetables
Concentration	Fruit and vegetable juices
Chemical preservation	Fruit semi-processed
Preservation with sugar	Fruit products/preserves
Pasteurization	Fruit and vegetable juices
Sterilisation	Fruits, vegetables
Sterilising filtration	Fruit juices
Irradiation	Fruits, vegetables

These preservation procedures have two main characteristics as far as being applied to all food products is concerned:

- Some of them are applied only to one or some categories of foods; others can be used across the board and thus a wider application (cold storage, freezing, drying/dehydration, sterilisation, etc.);
- Some guarantee food preservation on their own while others require combination with other procedures, either as principal or as auxiliary processes in order to assure preservation (for example smoking has to be preceded by salting).

Combined Preservation Procedures

In practice preservation procedures aim at avoiding microbiological and biochemical deterioration which are the principal forms of deterioration. Even with all recent progress achieved in this field, no single one of these technological procedures applied alone can be considered wholly satisfactory from a microbiological, physico-chemical and organoleptic point of view, even if to a great extent the food value is assured.

Thus, heat sterilisation cannot be applied in order to destroy all micro-organisms present in foods without inducing non desirable modifications. Preservation by dehydration/drying assures microbiological stability but has the drawback of undesirable modifications that appear during storage: vitamin losses, oxidation phenomena, etc.

Starting with these considerations, the actual tendency in food preservation is to study the application of combined preservation procedures, aiming at the realisation of maximum efficiency from a microbiological and biological point of view, with reduction to a minimum of organoleptical degradation and decrease in food value.

The principles of combined preservation procedures are:

- Avoid or reduce secondary (undesirable) effects in efficient procedures for microbiological preservation;
- Avoid qualitative degradation appearing during storage of products preserved by efficient procedures from a microbiological point of view;
- Increase microbiological efficiency of preservation procedures by supplementary means;
- Combine preservation procedures in order to obtain maximum efficiency from a microbiological point of view, by specific action on various types of micro-organisms present;
- Establish combined factors that act simultaneously on bacterial cells.

Research and applications in this direction were followed by microbiological and biochemical way, obtaining a serial of combination of preservation procedures with the possibility of application in industrial practice.

Fresh fruit and vegetable storage can be combined with:

- Storage in controlled atmosphere where carbon dioxide and oxygen levels are monitored, increasing concentration of CO2 and lowering that of oxygen according to fruit species. Excellent results were obtained for pomace fruit; in particular the storage period for apples has been extended. Application of this combined procedure requires airtight storage rooms.
- Storage in an environment containing ethylene oxide; this accelerates ripening in some fruit: tomatoes, bananas, mangoes, etc.

Cold storage can be combined with storage in an environment with added of carbon dioxide, sulphur dioxide, etc. according to the nature of product to be preserved.

Preservation by drying/dehydration can be combined with:- freezing: fresh fruit and vegetables are dehydrated up to the point where their weight is reduced by 50per cent and then they are preserved by freezing.

This procedure (freeze-drying) combines the advantages of drying (reduction of volume and weight) with those of freezing (maintaining vitamins and to a large extent organoleptic properties). A significant advantage of this process is the short drying time in so far as it is not necessary to go beyond the inflexion point of the drying curve. The finished products after defreezing and rehydration/reconstitution are of a better quality compared with products obtained by dehydration alone.

- Cold storage of dried/dehydrated vegetables in order to maintain vitamin C; storage temperature can be varied with storage time and can be at -8° C for a storage time of more than one year, with a relative humidity of 70-75 per cent.
- Packaging under vacuum or in inert gases in order to avoid action of atmospheric oxygen;, mainly for products containing beta-carotene.
- Chemical preservation: a process used intensively for prunes and which has commercial applications is to rehydrate the dried product up to 35 per cent using a bath containing hot 2 per cent potassium sorbate solution. Another possible application of this combined procedure is the initial dehydration up to 35per cent moisture followed by immersion in same bath as explained above; this has the advantage of reducing drying time and producing minimum qualitative degradation. Both applications suppress the dehydrated products reconstitution (rehydration) step before consumption.
- Packaging in the presence of desiccants (calcium oxide, anhydrous calcium chloride, etc.) in order to reduce water vapour content in the package, especially for powdered products.

Preservation by concentration, carried out by evaporation, is combined with cold storage during warm season for tomato paste (when water content cannot be reduced under the limit needed to inhibit moulds and yeasts, *e.g.* a_w = 0.70...0.75). Chemical preservation is combined with:

- Acidification of food medium (lowering pH);
- Using combined chemical preservatives.

Preservation by lactic fermentation (natural acidification) can be combined with cold storage for pickles in order to prolong storage time or shelf-life.

Preservation with sugar is combined with pasteurization for some preserves having a sugar content below 65per cent.

Fresh storage

Fresh Fruit and Vegetable Storage

Once fruit is harvested, any natural resistance to the action of spoiling micro-organisms is lost. Changes in enzymatic systems of the fruit also occur on harvest which may also accelerate the activity of spoilage organisms. Means that are commonly used to prevent spoilage of fruits must include:

- Care to prevent cutting or bruising of the fruit during picking or handling;
- Refrigeration to minimise growth of micro-organisms and reduce enzyme activity;

- Packaging or storage to control respiration rate and ripening;
- Use of preservatives to kill micro-organisms on the fruit.

A principal economic loss occurring during transportation and/or storage of produce such as fresh fruit is the degradation which occurs between the field and the ultimate destination due to the effect of respiration. Methods to reduce such degradation are as follows:

- Refrigerate the produce to reduce the rate of respiration;
- Vacuum cooling;
- Reduce the oxygen content of the environment in which the produce is kept to a value not above 5per cent of the atmosphere but above the value at which anaerobic respiration would begin. When the oxygen concentration is reduced within 60 minutes the deterioration is in practice negligible.

The following is a summary of some recent developments in post-harvest technology of fresh fruits and vegetables.

Harvest Maturity

This is particularly important with fruit for export. One recent innovation is the measurement of resonant frequency of the fruit which should enable the grading out of over mature and under-mature fruit before they are packed for export.

Harvest Method

Considerable research is continuing on mechanical harvesting of perishable crops with a view to minimising damage. In fruit trees, controlling their height by use of dwarfing rootstocks, pruning and growth regulating chemicals will lead to easier, cheaper more accurate harvesting.

Handling System

Field packing of various vegetables for export has been carried out for many years. In the last decade or so this has been applied, in selected cases, to a few tropical fruit types. Where this system can be practiced it has considerable economic advantages in saving the cost of building, labour and equipment and can result in lower levels of damage into crops.

Pre-cooling

Little innovation has occurred in crop pre-cooling over the last decade. However high velocity, high humidity forced air systems have continued to be developed and refined. These are suitable for all types of produce and are relatively simple to build and operate and, while not providing the speed of cooling of a vacuum or hydrocooler, have the flexibility to be used with almost all crops.

Chemicals

There is a very strong health lobby whose objective is to reduce the use of chemicals in agriculture and particularly during the post harvest period. Every year sees the prohibition of the use of commonly used post-harvest chemicals. New ways need to be developed to control post-harvest diseases, pest and sprouting.

Coatings

Slowing down the metabolism of fruit and vegetables by coating them with a material which affects their gaseous exchange is being tested and used commercially on a number of products.

Controlled Environment Transport

Recent innovations in this technique have produced great progress as a result of the development and miniaturisation of equipment to measure carbon dioxide and oxygen. Several companies now offer containers where the levels of these two gases can be controlled very precisely.

Water and Water Activity (a_w) in Foods

Micro-organisms in a healthy growing state may contain in excess of 80per cent water. They get this water from the food in which they grow. If the water is removed from the food it also will transfer out of the bacterial cell and multiplication will stop. Partial drying will be less effective than total drying, though for some micro-organisms partial drying may be quite sufficient to arrest bacterial growth and multiplication. Bacteria and yeasts generally require more moisture than moulds, and so moulds often will be found growing on semi-dry foods where bacteria and yeasts find conditions unfavourable; example are moulds growing on partially dried fruits.

Slight differences in relative humidity in the environment in which the food is kept or in the food package can make great differences in the rate of micro-organism multiplication. Since micro-organisms can live in one part of a food that may differ in moisture and other physical and chemical conditions from the food just millimetres away, we must be concerned with conditions in the "microenvironment" of the micro-organisms. Thus it is common to refer to water conditions in terms of specific activity.

The term "water activity" is related to relative humidity. Relative humidity is defined as the ratio of the partial pressure of water vapour in the air to the vapour pressure of pure water at the same temperature. Relative humidity refers to the atmosphere surrounding a material or solution. Water activity or aw is a property of solutions and is the ratio of vapour pressure of the solution compared with the vapour pressure of pure water at the same temperature. Under equilibrium conditions water activity equals:

$$a_w = RH/100$$

When we speak of moisture requirements of micro-organisms we really mean water activity in their immediate environment, whether this be in solution, in a particle of food or at a surface in contact with the atmosphere.

At the usual temperatures permitting microbial growth, most bacteria require a water activity in the range of about 0.90 to 1.00.

Some yeasts and moulds grow slowly at a water activity down to as low as about 0.65.

Qualitatively, water activity is a measure of unbound, free water in a system available to support biological and chemical reactions. Water activity, not absolute water content, is what bacteria, enzymes and chemical reactants encounter and is affected by at the micro-environmental level in food materials.

Two foods with the same water content can have very different a_w values depending upon the degree to which water is free or otherwise bound to food constituents. Fig is a representative water absorption isotherm for a given food at a given temperature. It shows the final moisture content the food will have when it reaches moisture equilibrium with atmospheres of different relative humidities.

Thus, this food, at the temperature for which this absorption isotherm was established, will ultimately attain a moisture content of 20per cent at 75per cent RH (relative humidity). If this food was previously dehydrated to below 20per cent moisture and placed in an atmosphere of 75per cent RH, it would absorb moisture until it reached 20per cent. Conversely, if this food was moistened to greater than 20per cent water and then placed at 75per cent RH, it would lose moisture until it reached the equilibrium value of 20per cent.

Under such conditions some foods may reach moisture equilibrium in the very short time of a few hours, others may require days or even weeks. When a food is in moisture equilibrium with its environment, then the a_w of the food will be quantitatively equal to the RH divided by 100.

Qualitatively, water activity is a measure of free or available water, to be distinguished from unavailable or bound water. These states of water also bear a relationship to the characteristic sigmoid shapes of water absorption isotherm curves of various foods.

Thus, according to theory, most of the water corresponding to the portion of the curve below its first inflection point is believed to be tightly bound water, often referred to as an adsorbed mono-molecular layer of water. Moisture corresponding to the region above this point and up to the curve's second inflection point is thought to exist largely as multi-molecular layers of water less tightly held to food constituent surfaces.

Beyond this second inflection point moisture generally is considered to be largely free water condensed in capillaries and interstices within the food. In this latter portion of the sorption isotherm curve small changes in moisture content result in great changes in a food's a_w. In Fig. given below are illustrated the moisture sorption isotherms for various dried fruits at 25°C.

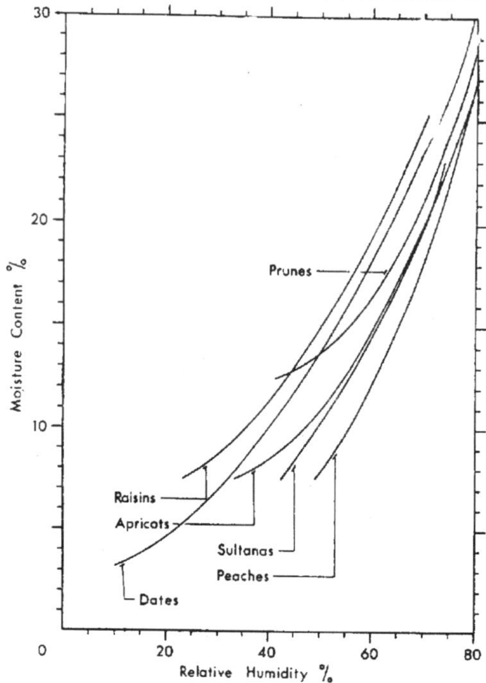

Fig. Moisture sorption isotherms for various dried fruits at 25°C.

Preservation by Drying/Dehydration

The technique of drying is probably the oldest method of food preservation practiced by mankind. The removal of moisture prevents the growth and reproduction of micro-organisms causing decay and minimises many of the moisture mediated deterioration reactions.

It brings about substantial reduction in weight and volume minimising packing, storage and transportation costs and enable storability of the product under ambient temperatures, features especially important for developing countries.

The sharp rise in energy costs has promoted a dramatic upsurge in interest in drying world-wide over the last decade.

Heat and Mass Transfer

Dehydration involves the application of heat to vapourise water and some means of removing water vapour after its separation from the fruit/vegetable tissues.

Hence it is a combined/simultaneous (heat and mass) transfer operation for which energy must be supplied. A current of air is the most common medium for transferring heat to a drying tissue and convection is mainly involved. The two important aspects of mass transfer are:

- The transfer of water to the surface of material being dried and
- The removal of water vapour from the surface.

In order to assure products of high quality at a reasonable cost, dehydration must occur fairly rapidly. Four main factors affect the rate and total drying time:

- The properties of the products, especially particle size and geometry;
- The geometrical arrangement of the products in relation to heat transfer medium (drying air);
- The physical properties of drying medium/environment;
- The characteristics of the drying equipment.

It is generally observed with many products that the initial rate of drying is constant and then decreases, sometimes at two different rates. The drying curve is divided into the constant rate period and the falling rate period.

Surface Area

Generally the fruit and vegetables to be dehydrated are cut into small pieces or thin layers to speed heat and mass transfer. Subdivision speeds drying for two reasons:

- Large surface areas provide more surface in contact with the heating medium (air) and more surface from which moisture can escape;
- Smaller particles or thinner layers reduce the distance heat must travel to the centre of the food and reduce the distance through which moisture in the centre of the food must travel to reach the surface and escape.

Temperature

The greater the temperature difference between the heating medium and the food the greater will be the rate of heat transfer into the food, which provides the driving force for moisture removal. When the heating medium is air, temperature plays a second important role. As water is driven from the food in the form of water vapour it must be carried away, or else the moisture will create a saturated atmosphere at the food's surface which will slow down the rate of subsequent water removal. The hotter the air the more moisture it will hold before becoming saturated. Thus, high temperature air in the vicinity of the dehydrating food will take up the moisture being driven from the food to a greater extent than will cooler air. Obviously, a greater volume of air also can take up more moisture than a lesser volume of air.

Air Velocity

Not only will heated air take up more moisture than cool air, but air in motion will be still more effective. Air in motion, that is, high velocity air, in addition to taking up moisture will sweep it away from the drying food's surface, preventing the moisture from creating a saturated atmosphere which would slow down subsequent moisture removal. This is why clothes dry more rapidly on a windy day. Some other phenomena influence the drying process and a few elements are summarised below.

Dryness of Air

When air is the drying medium of food, the drier the air the more rapid is the rate of drying. Dry air is capable of absorbing and holding moisture. Moist air is closer to saturation and so can absorb and hold less additional moisture than if it were dry. But the dryness of the air also determines how low a moisture content the food product can be dried to.

Atmospheric Pressure and Vacuum

If food is placed in a heated vacuum chamber the moisture can be removed from the food at a lower temperature than without a vacuum. Alternatively, for a given temperature, with or without vacuum, the rate of water removal from the food will be greater in the vacuum. Lower drying temperatures and shorter drying times are especially important in the case of heat-sensitive foods.

Evaporation and Temperature

As water evaporates from a surface it cools the surface. The cooling is largely the result of absorption by the water of the latent heat of phase change from liquid to gas. In doing this the heat is taken from the drying air or the heating surface and from the hot food, and so the food piece or droplet is cooled.

Time and Temperature

Since all important methods of food dehydration employ heat, and food constituents are sensitive to heat, compromises must be made between maximum possible drying rate and maintenance of food quality. As is the case in the use of heat for pasteurization and sterilisation, with few exceptions drying processes which employ high temperatures for short times do less damage to food than drying processes employing lower temperatures for longer times.

Thus, vegetable pieces dried in a properly designed oven in four hours would retain greater quality than the same products sun dried over two days.

Several drying processes will achieve dehydration in a matter of minutes or even less if the food is sufficiently subdivided.

Drying Techniques

Several types of dryers and drying methods, each better suited for a particular situation, are commercially used to remove moisture from a wide variety of food products including fruit and vegetables.

While sun drying of fruit crops is still practiced for certain fruit such as prunes, figs, apricots, grapes and dates, atmospheric dehydration processes are used for apples, prunes, and several vegetables; continuous processes as tunnel, belt trough, fluidised bed and foam-mat drying are mainly used for vegetables.

Spray drying is suitable for fruit juice concentrates and vacuum dehydration processes are useful for low moisture/high sugar fruits like peaches, pears and apricots. Factors on which the selection of a particular dryer/drying method depends include:

- Form of raw material and its properties;
- Desired physical form and characteristics of dried product;
- Necessary operating conditions;
- Operation costs.

There are three basic types of drying process:

- Sun drying and solar drying;
- Atmospheric drying including batch (kiln, tower and cabinet dryers) and continuous (tunnel, belt, belt-trough, fluidised bed, explosion puff, foam-mat, spray, drum and microwave);
- Sub-atmospheric dehydration (vacuum shelf/belt/drum and freeze dryers).

The scope has been expanded to include use of low temperature, low energy process like osmotic dehydration. As far dryers are concerned, one useful division of dryer types separates them into air convection dryers, drum or roller dryers, and vacuum dryers. Using this breakdown, Table given below indicates the applicability of the more common dryer types to liquid and solid type foods.

Table. Common Dryer Types Used for Liquid and Solid Foods.

Dryer type	Usual food type
Air convection dryers	
kiln	pieces
cabinet, tray or pan	pieces, purées, liquids
tunnel	pieces
continuous conveyor belt	purées, liquids
belt trough	pieces
air lift	small pieces, granules
fluidized bed	small pieces, granules
spray	liquid, purées
Drum or roller dryers	
atmospheric	purées, liquids
vacuum	purées, liquids
Vacuum dryers	
vacuum shelf	pieces, purées, liquids
vacuum belt	purées, liquids
freeze dryers	pieces, liquids

Fruit and Vegetable Natural Drying — Sun and Solar Drying

Surplus production and specifically grown crops may be preserved by natural drying for use until the next crop can be grown and harvested. Natural dried products can also be transported cheaply for distribution to areas where there are permanent shortages of fruit and vegetables.

The methods of producing sun and solar dried fruit and vegetables described here are simple to carry out and inexpensive. They can be easily employed by grower, farmer, cooperative, etc.

The best time to preserve fruits and vegetables is when there is a surplus of the product and when it is difficult to transport fresh materials to other markets. This is especially true for crops which are very easily damaged in transport and which stay in good condition for a very short time. Preservation extends the storage (shelf) life of perishable foods so that they can be available throughout the year despite their short harvesting season.

Sun and solar drying of fruits and vegetables is a cheap method of preservation because it uses the natural resource/source of heat: sunlight. This method can be used on a commercial scale as well at the village level provided that the climate is hot, relatively dry and free of rainfall during and immediately after the normal harvesting period. The fresh crop should be of good quality and as ripe (mature) as it would need to be if it was going to be used fresh. Poor quality produce cannot be used for natural drying.

Dried fruit and vegetables have certain advantages over those preserved by other methods. They are lighter in weight than their corresponding fresh produce and, at the same time, they do not require refrigerated storage. However, if they are kept at high temperatures and have a high moisture content they will turn brown after relatively short periods of storage.

Different lots at various stages of maturity (ripeness) must NOT be mixed together; this would result in a poor dried product. Some varieties of fruit and vegetables are better for natural drying than other; they must be able to withstand natural drying without their texture becoming tough so that they are not difficult to reconstitute. Some varieties are unsuitable because they have irregular shape and there is a lot of wastage in trimming and cutting such varieties.

Damaged parts which have been attacked by insects, rodents, diseases, etc. and parts which have been discoloured or have a bad appearance or colour, must be removed. Before trimming and cutting, most fruit and vegetables must be washed in clean water. Onions are washed after they have been peeled.

Trimming includes the selection of the parts which are to be dried, cutting off and disposing of all unwanted material. After trimming, the greater part of the fruit and vegetables cut into even slices of about 3 to 7 mm thick or in halves/quarters, etc.

It is very important to have all slices/parts in one drying lot of the same thickness/size; the actual thickness will depend on the kind of material. Uneven slices or different sizes dry at different rates and this result in a poor quality end product. Onions and root crops are sliced with a hand slicer or vegetable cutter; bananas, tomatoes and other vegetables or fruit are sliced with stainless-steel knives.

As a general rule plums, grapes, figs, dates are dried as whole fruits without cutting/slicing.

Some fruit and vegetables, in particular bananas, apples and potatoes, go brown very quickly when left in the air after peeling or slicing; this discolouration is due to an active enzyme called phenoloxidase. To prevent the slices from going brown they must be kept under water until drying can be started. Salt or sulphites in solution give better protection. However, whichever method is used, further processing should follow as soon as possible after cutting or slicing.

Blanching - exposing fruit and vegetable to hot or boiling water - as a pre-treatment before drying has the following advantages:

- It helps clean the material and reduce the amount of micro-organisms present on the surface;
- It preserves the natural colour in the dried products; for example, the carotenoid (orange and yellow) pigments dissolve in small intracellular oil drops during blanching and in this way they are protected from oxidative breakdown during drying;
- It shortens the soaking and/or cooking time during reconstitution.

During hot water blanching, some soluble constituents are leached out: water-soluble flavours, vitamins (vitamin C) and sugars. With potatoes this may be an advantage as leaching out of sugars makes the potatoes less prone to turning brown.

Blanching is a delicate processing step; time, temperature and the other conditions must be carefully monitored.

A suitable water-blanching method in traditional processing is as follows:

- The sliced material is placed on a square piece of clean cloth; the corners of the cloth are tied together;
- A stick is put through the tied corners of the cloth;
- The cloth is dipped into a pan containing boiling water and the stick rests across the top of the pan thus providing support for the cloth bag.

The average blanching time is 6 minutes. The start of blanching has to be timed from the moment the water starts to boil again after the cloth bag has been dipped into the pan. While the material is being blanched the cloth bag should be raised and lowered in the water so that the material is heated evenly.

When the blanching time is completed the cloth bag and its content should be dipped into cold water to prevent over-blanching. If products are over-blanched (boiled for too long) they will stick together on the drying trays and they are likely to have a poor flavour.

Green beans, carrots, okra, turnip and cabbage should always be blanched. The producer can choose whether or not potatoes need blanching. Blanching is not needed for onions, leeks, tomatoes and sweet peppers. Tomatoes are dipped into hot water for one minute when they need to be peeled but this is not blanching.

Use of Preservatives

Preservatives are used to improve the colour and keeping qualities of the final product for some fruits and vegetables. Preservatives include items such as sulphur dioxide, ascorbic acid, citric acid, salt and sugar and can either be simple or compound solutions.

Treatment with preservatives takes place after blanching or, when blanching is not needed, after slicing. In traditional, simple processing the method recommended is:

- Put enough preservative solution to cover the cloth bag into a container/pan;
- Dip the bag containing the product into the preservative solution for the amount of time specified;
- Remove the bag and put it on a clean tray while the liquid drains out. The liquid which drains out must not go back into the preservative solution because it would weaken the solution.

Care must be taken after each dip to refill the container to the original level with fresh preservative solution of correct strength. After five lots of material have been dipped, the remaining solution is thrown away; *i.e.* a fresh lot of preservative solution is needed for every 5 lots of material. The composition and strength of the preservative solution vary for different fruit and vegetables.

Table. Dilutions of sodium metabisulphite with water to obtain "PM" of sulphur dioxide (SO2).

PPM SO_2	SODIUM METABISULPHITE Grams per litre of water	Grams per 20 litre tin of water
1000	1.5	30.0
2000	3.0	60.0
3000	4.5	90.0
4000	6.0	120.0
5000	7.5	150.0
6000	9.0	180.0
7000	10.5	210.0

Note:

One level teaspoon of sodium metabisulphite = c. 5 g.

The strength of sulphur dioxide is expressed as "parts per million" (ppm). 1.5 grams of sodium metabisulphite in one litre of water gives 1000 ppm of sulphur dioxide. Details for solutions of different strengths are given in the following table.

The preservative solutions in the fruit and vegetable pre-treatment can only be used in enamelled, plastic or stainless-steel containers; never use ordinary metal because solutions will corrode this type of container.

As a general rule, preservatives are not used for treating onions, garlic, leeks, chilies and herbs.

Osmotic Dehydration

In osmotic dehydration the prepared fresh material is soaked in a heavy (thick liquid sugar solution) and/or a strong salt solution and then the material is sun or solar dried. During osmotic treatment the material loses some of its moisture. The syrup or salt solution has a protective effect on colour, flavour and texture.

This protective effect remains throughout the drying process and makes it possible to produce dried products of high quality. This process makes little use of sulphur dioxide.

Sun Drying

The main problems for sun drying are dust, rain and cloudy weather. Therefore, drying areas should be dust-free and whenever there is a threat of a dust storm or rain, the drying trays should be stacked together and placed under cover.

In order to produce dust-free and hygienically clean products, fruit and vegetable material should be dried well above ground level so that they are not contaminated by dust, insects, livestock or people. All materials should be dried on trays designed for the purpose; the most common drying trays have wooden frames with a fitted base of nylon mosquito netting. Mesh made of woven grass can also be used. Metal netting must NOT be used because it discolours the product.

The trays should be placed on a framework at table height from the ground. This allows the air to circulate freely around the drying material and it also keeps the food product well away from dirt. Ideally the area should be exposed to wind and this speed up drying, but this can only be done if the wind is free of dust.

With 80 cm x 50 cm trays, the approximate load for a tray is 3 kg; the material should be spread in even layers. During the first part of the drying period, the material should be stirred and turned over at least once an hour.

This will help the material dry faster and more evenly, prevent it sticking together and improve the quality of the finished product. Products for sun drying should be prepared early in the day; this will ensure that the material enjoys the full effect of the sun during the early stages of drying.

At night the trays should be stacked in a ventilated room or covered with canvas. Plastic sheets should NEVER be used for covering individual trays during sun drying. Dry or nearly dry products can be blown out of the tray by the wind. However, this can be protected by covering the loaded tray with an empty one; this also gives protection against insects and birds.

Shade Drying

Shade drying is carried out for products which can lose their colour and/or turn brown if put in direct sunlight. Products which have naturally vivid colours like herbs, green and red sweet peppers, chilies, green beans and okra give a more attractive end-product when they are dried in the shade.

The principles for the shade drying are the same as for sun drying. The material to be dried requires full air circulation. Therefore, shade drying is carried out under a roof or thatch which has open sides; it cannot be done either inside conventional buildings with side walls or in compounds sheltered from wind. Under dry conditions when there is a good circulation of air, shade drying takes little more time than is normally required for drying in full sunlight.

Identification of Suitable Designs of Solar Dryers for Different Applications

In the selection of appropriate solar dryers for commercial scale operation, it is imperative that economics be kept in view at all time.

A total Energy System concept should be employed and due consideration be given to parasitic energy consumption. The following features have been identified:

- Large scale dryers are more promising than small scale ones. However, small scale dryers should not be neglected.
- The dryer should be designed to maximise the utilisation factor of the capital investment, *i.e.* multi-products (fruit, vegetables and other raw material) and multi-use (*e.g.* drying and heating water for domestic use).
- In general, an auxiliary heat source should be provided to assure reliability, to handle peak loads and also to provide continuous drying during periods of no sunshine.
- Forced convection indirect dryers are preferred because they offer better control, more uniform drying and because of their high heat collection efficiency result in smaller collector area. However, parasitic power should be kept to a minimum.

Two dryer systems have been identified:

- A cabinet type dryer with natural convection for internal air circulation for the processing of dried fruit such as mango, banana, pineapple, apricot, pear, apple, etc. and also for potato chips and other vegetables;
- A greenhouse type dryer with forced air circulation.

Some of the barriers to the commercial development of solar dryers have been attributed to:

- Initial cost - poor farmers cannot afford them

- Durability - constant breakdown due to using low cost building materials
- Misuse - through lack of training and technical skills
- Dependability and reliability - during the wet season when drying is critical there is not enough solar energy available
- The wider use of Solar Drying Systems has been limited by other factors which are not necessarily of a technical or technological nature. Among the most important are the lack of national policies directed to promoting the drying of produce at the production site, in order to reduce losses, improve quality and increase farmers' earnings.

Construction of Solar Dryers

In the case of simple natural convection dryers it may be more appropriate to build and operate a number of small units. Multiplicity allows diversity, since more than one crop can be dried at a time. A further advantage is that if one dryer is out of operation due to damage, drying can still continue at reduced capacity using the other dryers. On the other hand, more sophisticated dryers, such as forced convection solar dryers, benefit from economies of scale due to the investment tied up in the fan and the source of heat.

Generally speaking, one large dryer will be more cost-effective than two smaller units. However, it should be taken into consideration that an oversized unit will be operating at less than full capacity, reducing any cost advantage. The drying area required will depend on local conditions, commodity and number of trays on each rack or trolley.

Construction Methods and Materials

Construction methods and available materials may vary considerably from location to location. It is not within the scope of this document to discuss individual, local circumstances. Some general guidelines regarding factors which must be considered can, however, be given:

- *Dimensions of standard materials:* Where possible, design should take account of the sizes of material locally available. For example, it would be poor design to specify the width of a corrugated iron collector as 1.1 m if the standard width of a corrugated iron sheet is 1 m.

Before finalising a design the commercial availability of materials must be ascertained.

- *Use of rural materials:* The cost of building of solar dryer can be minimised if the producer is able to use wood cut straight from the forest rather than prepared timber.

Careful design in the development stage of a dryer can often facilitate the use of cheaper materials. Difficulties caused by these materials are in joining pieces of the structure, in sealing the structure against air leaks, and in attaching the plastic sheet to the (wooden) frame. There is obvious scope for designs which use prepared timber for strategic points and unprepared at others.

Where the use of wood is necessary, remember to take environmental factors into consideration. For example, determine the effect of flash flooding or termites might be and take the appropriate preventive action.

- *Use of plastic sheets:* For many solar dryers, the clear plastic sheet used is the major capital cost to the farmer; therefore, the type of plastic chosen is important.

A choice must be made between a relatively cheap plastic such as ordinary polyethylene which will last, at best, for one season due to photo-degradation and wear and tear; and a more expensive, better quality plastic less prone to photo-degradation; or even glass or a rigid plastic.

Attaching plastic sheet to the framework structure, so as to minimise the likelihood of the plastic being torn is, perhaps, the most difficult part of building a dryer. Listed below are some general points which should be followed to prolong the useful lifetime of plastic sheet on a solar dryer:

- When attaching plastic sheet to the framework, care should be taken to stretch the plastic at the points of attachment, but the plastic should not be so loose that it will flap about in the wind;
- Rather than merely stapling or nailing the plastic directly to the framework, it is preferable to sandwich the plastic between the framework and a batten. This may not be practical when unprepared wood or other materials are being used.
- No sharp edges should come in contact with the plastic sheet since these will initiate tears;
- Fold over the plastic at the point of attachment to the frame, so that there are two or more layers of plastic. This will help prevent tears;
- When fixing the sheet over the framework, sags and hollows in which water can collect should be avoided wherever possible;
- The dryer should be handled as carefully and as seldom as possible during operation and when not in use.

Technical Criteria

The following design factors must be established:

- The throughput of the dryer over the productive season; the size of batch to be dried;
- The drying period(s) under stated conditions;
- The initial and desired final moisture content of the commodity (if known);
- The drying characteristics of the commodity, such as maximum drying temperature, effect of sunlight upon the product quality, etc.;
- Climatic conditions during the drying season, *i.e.* sunlight intensity

and duration; air temperature and humidity; wind speed (such data may be available from local meteorological stations);

- Availability and reliability of electrical power;
- The availability, quality, durability and price of potential construction materials such as:
  - Glazing materials: glass, plastic sheet or film;
  - Wood (prepared or unprepared);
  - Nails, screw, bolts, etc.;
  - Metal sheet, flat or corrugated angle iron;
  - Bricks (burnt or mud), concrete blocks, stones, cement, sand, etc.
  - Roofing thatch;
  - Metal mesh, wire netting, etc.
  - Mosquito netting, muslin, etc.
  - Bamboo or fibre weave;
  - Black paint, other blackening materials;
  - Insulation material; sawdust, etc.;
- The type of labour available to build and operate the dryer;
- The availability of clean water at the site for preparation of the commodity prior to drying.

In any one situation there may well be other technical factors that need to be considered.

Socio-economic Criteria

From the initial considerations, estimates of the capital costs of the dryer, the price of the commodity to be dried, and the likely selling price of the dried product will have been made. Other question that need to be considered are the following:

- Who will own the dryer?
- Is the dryer to be constructed by the end-user (with or without advice from extension agencies), local contractors, or other organisations?
- Who will operate and maintain it?
- How can the drying operation be incorporated into current practices?
- Are sources of finance from local authorities or extension agencies available, etc.?

Obviously there are many other socio-economic factors, particularly those of a local nature, which must be taken into account.

It cannot be stressed too highly that if such factors are not taken into account and evaluated, then is every chance that an inappropriate dryer design may result. Equal emphasis must be placed on both technical and socio-economic factors.

- Situations where solar dryer may be useful:

  - Where the cost of conventional energy is prohibitive and/or the supply is erratic, to supplement existing artificial drying systems and reduce fuel costs;
  - Where land is in short supply or expensive;
  - Where the quality of existing sun dried products can be improved upon;
  - Where the labour is in short supply;
  - Where is plenty of sunshine, but high humidity.
- Situations where solar dryers may not be useful:
  - Where conventional energy sources are abundant and cheap;
  - Where large amounts of combustible by-products or waste materials are freely available;
  - Where there is insufficient sunshine;
  - Where is plenty of sunshine and arid conditions (sun drying may suffice);
  - Where the quality of sun dried products already made cannot be improved upon;
  - Where local operators are insufficiently trained;
  - Where the ramifications of introducing a solar dryer have not been completely thought out.

Sun/solar Drying Tray

The drying tray described requires seasoned timber 22.5 mm thick x 50 mm wide. A sun drying tray requires 6 metres of seasoned timber 22.5 mm thick x 50 mm wide. Cut the timber into lengths of 900 mm long for the sides of the tray and 600 mm long for the ends - 4 pieces of each length will be needed. The ends of each piece are cut as shown in the drawing - this is to make flush fitting joints. Join the corners using small brass screws 20 mm long. To make extra strong joints use good quality wood glue as well as the screws. The nylon mosquito netting or grass woven mesh can be fitted between the frames as shown in the bottom drawing. Cut the mesh a little larger than the size of the frame. Using drawing pins, pin the mesh to the OUTSIDE edges of one of the frames - the mesh should be pulled tight as the pins are put in around the edges. Lay the other frame on top and drill holes about 3 mm in diametre at the points marked X in the top drawing. Use nails that are a tight fit in the holes and tap gently into place leaving a portion standing above the frame. Cut off the standing part to leave a piece about 12 mm long which is then bent over and tapped firmly down onto the frame. When the frame has been put together tightly, the drawing pins can be removed.

10

Food Safety and Security

IMPORTANT FOOD ISSUES

FOOD SAFETY, QUALITY AND CONSUMER PROTECTION

The terms food safety and food quality can sometimes be confusing. Food safety refers to all those hazards, whether chronic or acute, that may make food injurious to the health of the consumer. It is not negotiable. Quality includes all other attributes that influence a product's value to the consumer. This includes negative attributes such as spoilage, contamination with filth, discoloration, off-odours and positive attributes such as the origin, colour, flavour, texture and processing method of the food. This distinction between safety and quality has implications for public policy and influences the nature and content of the food control system most suited to meet predetermined national objectives. Food control is defined as:

- A mandatory regulatory activity of enforcement by national or local authorities to provide consumer protection and ensure that all foods during production, handling, storage, processing, and distribution are safe, wholesome and fit for human consumption; conform to safety and quality requirements; and are honestly and accurately labelled as prescribed by law.

The foremost responsibility of food control is to enforce the food law(s) protecting the consumer against unsafe, impure and fraudulently presented food by prohibiting the sale of food not of the nature, substance or quality demanded by the purchaser. Confidence in the safety and integrity of the food supply is an important requirement for consumers. Foodborne disease outbreaks involving agents such as *Escherichia coli*, Salmonella and chemical contaminants highlight problems with food safety and increase public anxiety that modern farming systems, food processing and marketing do not provide adequate safeguards for public health. Factors which contribute to potential hazards in foods include improper agricultural practices; poor hygiene at all stages of the food chain;

lack of preventive controls in food processing and preparation operations; misuse of chemicals; contaminated raw materials, ingredients and water; inadequate or improper storage, etc. Specific concerns about food hazards have usually focused on:

- Microbiological hazards;
- Pesticide residues;
- Misuse of food additives;
- Chemical contaminants, including biological toxins; and
- Adulteration.

The list has been further extended to cover genetically modified organisms, allergens, veterinary drugs residues and growth promoting hormones used in the production of animal products. Consumers expect protection from hazards occurring along the entire food chain, from primary producer through consumer.

Protection will only occur if all sectors in the chain operate in an integrated way, and food control systems address all stages of this chain. As no mandatory activity of this nature can achieve its objectives fully without the cooperation and active participation of all stakeholders *e.g.* farmers, industry, and consumers, the term Food Control System is used in these Guidelines to describe the integration of a mandatory regulatory approach with preventive and educational strategies that protect the whole food chain.

Thus an ideal food control system should include effective enforcement of mandatory requirements, along with training and education, community outreach programmes and promotion of voluntary compliance. The introduction of preventive approaches such as the Hazard Analysis Critical Control Point System, have resulted in industry taking greater responsibility for and control of food safety risks. Such an integrated approach facilitates improved consumer protection, effectively stimulates agriculture and the food processing industry, and promotes domestic and international food trade.

GLOBAL CONSIDERATIONS

International Trade

With an expanding world economy, liberalization of food trade, growing consumer demand, developments in food science and technology, and improvements in transport and communication, international trade in fresh and processed food will continue to increase. Access of countries to food export markets will continue to depend on their capacity to meet the regulatory requirements of importing countries. Creating and sustaining demand for their food products in world markets relies on building the trust and confidence of importers and consumers in the integrity of their food systems. With agricultural production the focal point of the economies of most developing countries, such food protection measures are essential.

Codex Alimentarius Commission

The Codex Alimentarius Commission is an intergovernmental body that coordinates food standards at the international level. Its main objectives are to protect the health of consume and ensure fair practices in food trade. The CAC has proved to be most successful in achieving international harmonization in food quality and safety requirements. It has formulated international standards for a wide range of food products and specific requirements covering pesticide residues, food additives, veterinary drug residues, hygiene, food contaminants, labelling, etc. These Codex recommendations are used by governments to determine and refine policies and programmes under their national food control system.

More recently, Codex has embarked on a series of activities based on risk assessment to address microbiological hazards in foods, an area previously unattended. Codex work has created worldwide awareness of food safety, quality and consumer protection issues, and has achieved international consensus on how to deal with them scientifically, through a risk-based approach. As a result, there has been a continuous appraisal of the principles of food safety and quality at the international level. There is increasing pressure for the adoption of these principles at the national level.

SPS and TBT Agreements

The conclusion of the Uruguay Round of Multilateral Trade Negotiations in Marrakech led to the establishment of the WTO on 1st January 1995, and to the coming into force of the Agreement on the Application of Sanitary and Phytosanitary Measures and the Agreement on Technical Barriers to Trade. Both these Agreements are relevant in understanding the requirements for food protection measures at the national level, and the rules under which food is traded internationally. The SPS Agreement confirms the right of WTO member countries to apply measures to protect human, animal and plant life and health. The Agreement covers all relevant laws, decrees, regulations; testing, inspection, certification and approval procedures; and packaging and labelling requirements directly related to food safety.

Member States are asked to apply only those measures for protection that are based on scientific principles, only to the extent necessary, and not in a manner which may constitute a disguised restriction on international trade. The Agreement encourages use of international standards, guidelines or recommendations where they exist, and identifies those from Codex to be consistent with provisions of SPS. Thus, the Codex standards serve as a benchmark for comparison of national sanitary and phytosanitary measures. While it is not compulsory for Member States to apply Codex Standards, it is in their best interests to harmonize their national food standards with those elabourated by Codex. The TBT Agreement requires that technical regulations on traditional quality factors, fraudulent practices, packaging, labelling, etc., imposed by countries will not be more restrictive on imported products than they are on products produced domestically. It also encourages use of international standards.

ELEMENTS OF A NATIONAL FOOD CONTROL SYSTEM

OBJECTIVES

The principal objectives of national food control systems are:

- Protecting public health by reducing the risk of foodborne illness;
- Protecting consumers from unsanitary, unwholesome, mislabelled or adulterated food; and
- Contributing to economic development by maintaining consumer confidence in the food system and providing a sound regulatory foundation for domestic and international trade in food.

SCOPE

Food control systems should cover all food produced, processed and marketed within the country, including imported food. Such systems should have a statutory basis and be mandatory in nature.

BUILDING BLOCKS

While the components and priorities of a food control system will vary from country to country, most systems will typically comprise the following components:

Food Law and Regulations

The development of relevant and enforceable food laws and regulations is an essential component of a modern food control system. Many countries have inadequate food legislation and this will impact on the effectiveness of all food control activities carried out in the country. Food law has traditionally consisted of legal definitions of unsafe food, and the prescription of enforcement tools for removing unsafe food from commerce and punishing responsible parties after the fact. It has generally not provided food control agencies with a clear mandate and authority to prevent food safety problems. The result has been food safety programmes that are reactive and enforcement-oriented rather than preventive and holistic in their approach to reducing the risk of foodborne illness. To the extent possible, modern food laws not only contain the necessary legal powers and prescriptions to ensure food safety, but also allow the competent food authority or authorities to build preventive approaches into the system. In addition to legislation, governments need updated food standards. In recent years, many highly prescriptive standards have been replaced by horizontal standards that address the broad issues involved in achieving food safety objectives. While horizontal standards are a viable approach to delivering food safety goals, they require a food chain that is highly controlled and supplied with good data on food safety risks and risk management strategies and as such may not be feasible for many developing countries.

Similarly, many standards on food quality issues have been cancelled and replaced by labelling requirements. In preparing food regulations and standards, countries should take full advantage of Codex standards and food safety sessions learned in other countries. Taking into account the experiences in other countries while tailoring the information, concepts and requirements to the national context is the only sure way to develop a modern regulatory framework that will both satisfy national needs and meet the demands of the SPS Agreement and trading partners.

Food legislation should include the following aspects:

- It must provide a high level of health protection;
- It should include clear definitions to increase consistency and legal security;
- It should be based on high quality, transparent, and independent scientific advice following risk assessment, risk management and risk communication;
- It should include provision for the use of precaution and the adoption of provisional measures where an unacceptable level of risk to health has been identified and where full risk assessment could not be performed;
- It should include provisions for the right of consumers to have access to accurate and sufficient information;
- It should provide for tracing of food products and for their recall in case of problems;
- It should include clear provisions indicating that primary responsibility for food safety and quality rests with producers and processors;
- It should include obligation to ensure that only safe and fairly presented food is placed on the market;
- It should also recognise the country's international obligations particularly in relation to trade; and
- It should ensure transparency in the development of food law and access to information.

Food Control Management

Effective food control systems require policy and operational coordination at the national level. While the detail of such functions will be determined by the national legislation, they would include the establishment of a leadership function and administrative structures with clearly defined accountability for issues such as: the development and implementation of an integrated national food control strategy; operation of a national food control programme; securing funds and allocating resources; setting standards and regulations; participation in international food control related activities; developing emergency response procedures; carrying out risk analysis; etc. Core responsibilities include the establishment of regulatory measures, monitoring system performance, facilitating continuous improvement, and providing overall policy guidance.

Inspection Services

The administration and implementation of food laws require a qualified, trained, efficient and honest food inspection service. The food inspector is the key functionary who has day-to-day contact with the food industry, trade and often the public. The reputation and integrity of the food control system depends, to a very large extent, on their integrity and skill. The responsibilities of the inspection services include:

- Inspecting premises and processes for compliance with hygienic and other requirements of standards and regulations;
- Evaluating HACCP plans and their implementation;
- Sampling food during harvest, processing, storage, transport, or sale to establish compliance, to contribute data for risk assessments and to identify offenders;
- Recognizing different forms of food decomposition by organoleptic assessment; identifying food which is unfit for human consumption; or food which is otherwise deceptively sold to the consumer; and taking the necessary remedial action;
- Recognizing, collecting and transmitting evidence when breaches of law occur, and appearing in court to assist prosecution;
- Encouraging voluntary compliance in particular by means of quality assurance procedures;
- Carrying out inspection, sampling and certification of food for import/ export inspection purposes when so required;
- In establishments working under safety assurance programmes such as HACCP, conduct risk based audits.

Proper training of food inspectors is a prerequisite for an efficient food control system. As current food systems are quite complex, the food inspector must be trained in food science and technology to understand the industrial processes, identify potential safety and quality problems, and have the skill and experience to inspect the premises, collect food samples and carry out an overall evaluation. The inspector must have a good understanding of the relevant food laws and regulations, their powers under those laws, and the obligations such laws impose on the food sector. They should also be conversant with procedures for collecting evidence, writing inspection reports, collecting samples and sending them to a labouratory for analysis. With gradual introduction of HACCP systems in the food industry, the inspector should be trained to handle HACCP audit responsibilities.

Clearly, there is a continuing need for training and upgrading the skills of existing inspectional staff and having a policy for human resource development, especially the development of inspectional specialists in specific technical areas. As human resources in some food control agencies in developing countries may be limited, environmental health inspectors are often also asked to work as food inspectors. This is not the ideal situation as they may lack the skills and knowledge to effectively evaluate and inspect food operations. If environmental health inspectors must be used, then they should be carefully supervised and provided with on-the-job training.

Labouratory Services: Food Monitoring and Epidemiological Data

Labouratories are an essential component of a food control system. The establishment of labouratories requires considerable capital investment and they are expensive to maintain and operate. Therefore careful planning is necessary to achieve optimum results. The number and location of the labouratories should be determined in relation to the objectives of the system and the volume of work.

If more than one labouratory is required, consideration should be given to apportioning the analytical work to achieve the most effective coverage of the food analyses to be performed and also to having a central reference labouratory equipped for sophisticated and reference analyses. All food analysis labouratories may not be under the control of one agency or ministry, and a number could be under the jurisdiction of the states, provinces and local authorities. The Food Control Management should, however, lay down the norms for food control labouratories and monitor their performance. The labouratories should have adequate facilities for physical, microbiological and chemical analyses. In addition to simple routine analysis, the labouratories can be equipped with more sophisticated instruments, apparatus and library facilities as required.

It is not only the type of equipment that determines the accuracy and reliability of analytical results but also the qualification and skill of the analyst and the reliability of the method used. The analytical results of a food control labouratory are often used as evidence in a court of law to determine compliance with regulations or standards of the country. It is therefore necessary that utmost care be taken to ensure the efficient and effective performance of the labouratory. The introduction of analytical quality assurance programmes and accreditation of the labouratory by an appropriate accreditation agency within the country or from outside, enables the labouratory to improve its performance and to ensure reliability, accuracy and repeatability of its results. Prescription of official methods of sampling and analysis also support this effort. An important element of a national food control system is its integration in a national food safety system so that links between food contamination and foodborne diseases can be established and analysed. Access to reliable and current intelligence on the incidence of foodborne illness is critical.

The labouratory facilities for this type of activity are generally situated outside the food control agencies. It is essential, however, that effective linkages are established between food control agencies and the public health system including epidemiologists and microbiologists. In this way information on foodborne diseases may be linked with food monitoring data, and lead to appropriate risk-based food control policies. This information includes annual incidence trends, identification of susceptible population groups, identification of hazardous foods, identification and tracing of causes of foodborne diseases, and the development of early warning systems for outbreaks and food contamination.

Information, Education, Communication and Training

An increasingly important role for food control systems is the delivery of information, education and advice to stakeholders across the farm-to-table continuum. These activities include the provision of balanced factual information to consumers; the provision of information packages and educational programmes for key officials and workers in the food industry; development of train-the-trainer programmes; and provision of reference literature to extension workers in the agriculture and health sectors. Food control agencies should address the specific training needs of their food inspectors and labouratory analysts as a high priority. These activities provide an important means of building food control expertise and skills in all interested parties, and thereby serve an essential preventive function.

CONSIDERATIONS OF FOOD SAFETY AND CONSUMER PROTECTION

Food safety is an essential public health issue for all countries. Foodborne diseases due to microbial pathogens, biotoxins, and chemical contaminants in food represent serious threats to the health of thousands of millions of people. Serious outbreaks of foodborne disease have been documented on every continent in the past decades, illustrating both the public health and social significance of these diseases.

Consumers everywhere view foodborne disease outbreaks with ever-increasing concern. Outbreaks are likely, however, to be only the most visible aspect of a much broader, more persistent problem. Foodborne diseases not only significantly affect people's health and well-being, but they also have economic consequences for individuals, families, communities, businesses and countries. These diseases impose a substantial burden on health care systems and markedly reduce economic productivity. Poor people tend to live from day-to-day, and loss of income due to foodborne illness perpetuates the cycle of poverty. The integration and consolidation of food industries and the globalization of the food trade are changing the patterns of food production and distribution. Food and feed are distributed over far greater distances than before, creating the conditions necessary for widespread outbreaks of foodborne illness. In a recent crisis, more than 1,500 farms in Europe received dioxincontaminated feed from a single source over a two-week period. Food produced from animals given this contaminated fodder found its way onto every continent within weeks.

The international spread of meat and bonemeal prepared from cattle affected by bovine spongiform encephalitis needs no further description. The full economic consequences of such incidents and the anxiety raised among consumers are still being assessed. Other factors account for the emergence of food safety as a public health issue. Increasing urbanization leads to greater requirements for transport, storage and preparation of food.

In developing countries, food is often prepared by street vendors. In developed countries, up to 50 per cent of the food budget may be spent on food prepared outside the home. All these changes lead to situations in which a single source of contamination can have widespread, even global conse-quences. The globalization of the food trade offers many benefits to consumers, as it results in a wider variety of high-quality foods that are accessible, affordable and safe, meeting consumer demand. The global food trade provides opportunities for food-exporting countries to earn foreign exchange, which is indispensable for the economic development of many countries.

However, these changes also present new challenges to safe food production and distribution and have been shown to have widespread repercussions on health. Food safety programmes are increasingly focusing on a farm-to-table approach as an effective means of reducing foodborne hazards. This holistic approach to the control of food-related risks involves consideration of every step in the chain, from raw material to food consumption. Hazards can enter the food chain on the farm and can continue to be introduced or exacerbated at any point in the chain. Although significant progress has been made in many countries in making food safer, thousands of millions of people become ill each year from eating contaminated food. The emergence of increased antimicrobial resistance in bacteria causing disease is aggravating this picture. The public is increasingly aware of the risks posed by pathogenic microorganisms and chemical substances in the food supply. The introduction of new technologies, including genetic engineering and irradiation, in this climate of concern about food safety is posing a special challenge. Some new technologies will increase agricultural production and make food safer, but their usefulness and safety must be demonstrated if they are to be accepted by consumers. Furthermore, the evaluation must be participatory, transparent and conducted using internationally agreed methods. Until recently, most systems for regulating food safety were based on legal definitions of unsafe food, enforcement programmes for the removal of unsafe food from the market and sanctions for the responsible parties after the fact.

These traditional systems cannot respond to existing and emerging challenges to food safety because they do not provide or stimulate a preventive approach. During the past decade, there was a transition to risk analysis based on better scientific knowledge of foodborne illness and its causes. This provides a preventive basis for regulatory measures for food safety at both national and international levels. The risk-based approach must be backed by information on the most appropriate and effective means to control foodborne hazards.

MICROBIOLOGICAL HAZARDS

The dangers of foodborne pathogenic microorganisms have been known for decades. The risk of milk-borne transmission of tuberculosis and salmonellosis was recognized early in the twentieth century, and control by pasteurization was an early intervention. Similarly, problems with botulism were managed by controlling the application of heat to low acid foods in hermetically sealed containers. Despite remarkable advances in food science and technology, foodborne illness is a rising cause of morbidity in all countries and the list of potential foodborne microbial pathogens keeps increasing.

Furthermore, foodborne illness is a major cause of preventable death and economic burden in most countries. Unfortunately, most countries have limited data on foodborne disease and its impact on public health. It is only recently that the burden of food contamination and foodborne disease have been systematically assessed and quantified. Studies on foodborne disease outbreaks in the USA, Australia, Germany, and India have confirmed the enormity of the problem with millions being affected or dying from foodborne diseases. The data indicates that up to 30 per cent of the population in industrialized countries may be affected by foodborne illness each year.

The global incidence of foodborne disease is difficult to estimate, but in 1998 it was estimated that 2.2 million people, including 1.8 million children, died from diarrhoeal diseases. The economic cost associated with foodborne diseases caused by microorganisms has only recently been estimated. In the US, the costs of human illness due to seven specific pathogens have been estimated to range between US $6.5 billion to $34.9 billion. The medical costs and value of lives lost from five foodborne infections in England and Wales were estimated at UK £300-700 million annually in 1996.

The cost of an estimated 11,500 cases of food poisoning per day in Australia was calculated at AUD $2.6 billion annually. Yet on the basis of per capita income, the economic burden on people in India affected by an outbreak of Staphylococcus aureus food poisoning was found to be higher than in case of a similar outbreak in the US. Major outbreaks involving *E.coli* and Salmonella have highlighted problems with food safety and increased public anxiety that modern farming systems, food processing and marketing may not provide adequate safeguards for public health. While our understanding of the ecology of food poisoning organisms and the environment in which they may grow and survive has increased, our ability to control some of these organisms has declined.

This may be due in part to modified production practices, lack of control of hazards at the farm level, industry difficulties in controlling hazards during production, growing demand for fresh foods, the trend towards minimal processing of foodstuffs and longer shelf-life for many foodstuffs. For example, Salmonella persists as a major cause of food poisoning and its incidence is on the rise. *Salmonella typhimurium* is widely distributed in cattle herds and is resistant to several antibiotics. The incidence of this organism is on the increase, as is the number of antibiotic resistant isolates.

More than one third of people infected by this organism require hospitalization with approximately three per cent of cases being fatal. Enterohaemorrhagic *Escherichia coli* has emerged as a significant foodborne pathogen, and has been highly publicised through major outbreaks of disease. E.coli 0157:H7 was first recognized as a pathogen in 1982, but progress to identify reservoirs and sources of the organism was initially hampered by a lack of sufficiently sensitive detection methods.

Other strains of enterohaemorrhagic *E.coli* pose a particular problem in that they are impossible to differentiate culturally from other flora in the gut. Identification requires advanced techniques. *E.coli* serves as an example of the limitation of the present knowledge and understanding of many pathogens and how they contaminate food. In the past few decades a range of microorganisms have emerged as potential causes of foodborne disease. Several relatively unknown bacteria have been identified as major causes of foodborne illness, *e.g. Campylobacter jejuni*, *Vibrio parahaemolyticus*, and *Yersinia enterocolitica* may cause foodborne illness. As microorganisms have the ability to adapt, modified modes of food production, preservation and packaging have resulted in altered food safety hazards. For example, organisms such as Listeria monocytogenes, and to a lesser extent *Clostridium botulinum*, have emerged and re-emerged because of changes in the way high-risk foods are packaged and processed. A range of protozoa and viruses may also infect food, *e.g. Cryptosporidium parvum*, *Toxoplasma gondii*, *Clonorchis sinensis*, *Norwalk virus*, and hepatitis A. The effective prevention and control of these organisms requires widespread education and possibly new initiatives such as HACCP being introduced at primary production level.

CHEMICAL HAZARDS

Chemical hazards are also a significant source of foodborne illness, although the effect is often difficult to link to a particular food and may occur long after consumption. In particular, there have been long-standing concerns about the chemical safety of food due to misuse of pesticides during food production and storage, resulting in the occurrence of undesirable residues.

Similarly, heavy metal contaminants can enter food through soil or water or food contact material, as can other environmental contaminants such as PCBs. All can lead to acute or chronic illness. More recently, contamination from dioxins entering the animal feed supply has highlighted both the importance of controlling the whole food chain and the international concerns about food safety systems.

Misuse and illegal use of food additives create their own food safety problems. Phthalates in infant formulae, substances in food with oestrogenic activity, and veterinary drug residues, etc have also heightened public concern. These problems are not confined to food produced on land. They also include algal toxins in fish and the widespread use of chemicals in fish farming. Mycotoxins are another group of highly toxic or carcinogenic chemical contaminants of biological origin produced by certain species of fungi. Five important mycotoxins are aflatoxins, ochratoxins, fumonisins, zearalenone, and trichothecenes. Crops such as peanuts, corn, pistachio, walnuts, copra are susceptible to mycotoxin contamination.

Aflatoxins are among the most studied mycotoxins, and the relationship between aflatoxin ingestion and primary liver cancer is well established. Almost all plant products can serve as substrates for fungal growth, and subsequently mycotoxin contamination of human food and animal feed. Animal feed contaminated with mycotoxins can result in the carry-over of toxins through milk and meat to consumers.

While the importance of chemicals hazards is well recognized, our understanding of the effect of chemicals in food intolerances and allergies, endocrine system disruption, immunotoxicity, and certain forms of cancer are incomplete. Further research is necessary to determine the role of chemicals in foods in the etiology of these diseases. In developing countries, little reliable information is available on the exposure of the population to chemicals in food.

FOOD ADULTERATION

Consumers, particularly in developing countries, are often exposed to wilful adulteration of their food supply. This can lead to health hazards and to financial losses for the consumer. Adulteration of milk and milk products, honey, spices, edible oils, and the use of colours to mask product quality to cheat the consumer are quite common. Although risks associated with adulteration are usually low, such episodes invoke public outrage and anger as it violates public trust in the integrity of the food supply. With 60-70 per cent of the income of middle class families in developing countries being spent on food, food adulteration can impact heavily on both the family budget and the health status of the family members.

GENETICALLY MODIFIED ORGANISMS AND NOVEL FOODS

Modern biotechnology, also referred to as genetic engineering or genetic manipulation, involves transfer of hereditary material from one organism to another in a way that cannot be achieved naturally *i.e.* through mating or cross breeding. Genetic engineering can now transfer the hereditary material across species boundaries. This can broaden the range of genetic changes that can be made to food and can expand the spectrum of possible food sources. The accelerating pace of developments in modern biotechnology has opened a new era in food production and this may have a tremendous impact on world food supply systems.

However, there are considerable differences of opinion among scientists about the safety, nutritional value and environmental effects of such foods. Overall, it is argued that the consequences of some gene transfer methods are less predictable when compared to those of traditional plant breeding methods and considerable scientific evidence will be needed to clear these foods from points of view of nutrition, food safety and impact on the environment.

The revolutionary nature of modern biotechnology and its likely impact on world food resources has created worldwide interest and debate among scientists, consumers and industry as well as among policy makers at national and international levels.

URBANIZATION, NUTRITION AND FOOD SECURITY

In 2020, the world population is projected to reach 7.6 billion, an increase of 31 per cent over the mid-1996 population of 5.8 billion. Approximately 98 per cent of the population growth occurring during this period will take place in developing countries. While urbanization is a global phenomenon, it has been estimated that between the years 1995 and 2020 the developing world's urban population will double, reaching 3.4 billion. Such population growth poses great challenges to world food security and food systems.

Further extension of improved agriculture and animal husbandry practices; use of measures to prevent and control pre- and post-harvest losses; more efficient food processing and distribution systems; introduction of new technologies including the application of biotechnology, and others will have to be exploited to increase food availability to meet the needs of growing populations. Growing urbanization and associated changes in the way food is produced and marketed have led to a lengthening of the food chain and potential for introducing or exacerbating foodborne hazards.

FOOD SAFETY CAPACITY DEVELOPMENT

In order to support countries wishing to fulfil the mandate set by the CPB, the secretariat of the CBD maintains a global database of capacity-building initiatives as a component of the Biosafety Clearing House. The purpose of this database is to give an overview of past, present and prospective capacity-building initiatives. The secretariat intends to use the information to develop a method for coordinating capacity-building initiatives, thus ensuring that they complement one another, use funding efficiently and strengthen resources in the recipient countries. Although the secretariat's interest is in initiatives that would support the effective implementation of the CPB, the database covers a wider range of initiatives, such as technology transfer and those directed towards biotechnology research. To date, 89 initiatives have been listed in the database, illustrating a broad range of implementing agencies.

The secretariat, more than half of the registered initiatives have been negotiated bilaterally and through industry interest groups. United Nations agencies, intergovernmental organizations, or individual governments, industry or NGOs supported most of these countries through bilateral agreements. While the capacity-building initiatives collectively cover all the aspects associated with the application of modern biotechnology, no single one covers the entire range—each is limited to its own specific focus. For example, the FAO/WHO expert consultations and capacity-building programmes supported by both organizations train individuals in food-safety-related issues only.

WHO has advised Member States and assisted in building their capacity for food-safety-related issues for many years. The food-safety activities of WHO have increased significantly over the years, with the establishment of international expert scientific bodies such as the Joint FAO/WHO Expert Committee on Food Additives in 1956, to evaluate the safety of food additives, contaminants, naturally occurring toxicants and residues of veterinary drugs in food; the Joint FAO/WHO Meeting on Pesticide Residues to evaluate the safety of pesticide residues in food; the Joint FAO/WHO Expert Meeting on Microbiological Risk Assessment to provide risk-assessment guidelines for selected pathogens and for microbiological hazards in food and water. Also in 1963, the Codex Alimentarius Commission was created to implement the joint FAO/WHO Food Standards Programme. To strengthen its in-house activities, WHO created the Food Safety Programme in 1978, operating at national, regional and international levels. The recognition of food safety as a major public-health concern by the World Health Assembly in 2000 has also increased the profile of food-safety-related issues, not only within the Organization but also at the national level. These activities were further supported by the endorsement of the WHO Global Strategy for Food Safety by the WHO Executive Board in 2002. In this strategy, WHO proposes to "formulate regional food safety strategies on the basis of the WHO global food safety strategy and of specific regional needs such as technical support, educational tools and training". Considerable technical assistance has been provided to developing countries to create and/or enhance food-safety control systems, but these activities have not been effectively coordinated and therefore not been adequate in meeting the public-health needs of recipient countries.

The SPS Agreement of WTO calls for assistance to developing-country Members to enable them to strengthen their food safety and animal and plant health protection. The Agreement encourages Members to enter into bilateral arrangements for technical assistance, or to seek training through other international organizations. Such assistance can be in the area of processing technology, research or infrastructure development, and may take the form of technical advice, expertise, financial assistance or procurement of adequate equipment. Food-safety activities within WHO take place at the international, regional and country levels. The regional and country offices provide assistance in developing and strengthening national food-safety programmes, whereas WHO headquarters develops guidelines for such work, including the framework for risk analyses and setting international standards. The division of these activities is arbitrary as headquarters also participates in activities at the national and regional levels, with technical know-how and capacity-building guidance.

These activities include:

- Developing regional and national food-safety policy and strategies;
- Preparing food legislation, regulations, standards and codes of hygienic practice;
- Implementing food inspection programmes;
- Promoting methods and technologies designed to prevent foodborne diseases, including the hazard analysis and critical control point system;
- Developing or enhancing food analysis capability;
- Developing methods for assessing the safety of the products of new technologies;
- Establishing healthy markets and enhancing the safety of street food; and
- Promoting the establishment of foodborne disease surveillance systems.

Many WHO activities to build food-safety capacity are developed in collabouration with FAO. However, FAO also administers a major, separate technical cooperation programme building capacity in this and other agriculture-related areas in many developing countries. Although most developing countries have national food-control systems, these are often not based on modern scientific concepts. Moreover, they cannot be adapted to cope with developments in food science and technology. The specifications for an effective food-control system include: regulations, capacity for assessing the risks associated with the food, and ongoing monitoring and evaluation of the risks. A capacity-building programme for the risk assessment of products of modern biotechnology would involve:

- Use of the concept of a comparative safety assessment;
- Hazard identification and characterization;
- Assessment of food intake, including consumption profile and effects;
- Use of integrated toxicological evaluation;
- Use of integrated nutritional evaluation;
- Risk characterization; and
- Application of risk-manageent strategies, such as labelling and monitoring.

Other Considerations

competent authorities need information relating to trends in biotechnology and biosafety to keep abreast with biotechnology developers. Information exchange systems as provided by a number of organizations fulfil this need by facilitating international cooperation, but can only be used by developing countries where the appropriate expertise exists. Even more limiting, several of these information networks are difficult to search, while others are limited in scope.

The Intergovernmental Committee for the CPB, realizing the capacity constraints of developing countries, has established a coordination mechanism to maximize synergies, complementarity and collabouration between the numerous international initiatives. The evaluation of food is not only about science. It should also take into account the social, ethical and religious concerns of the local populations.

HARMONIZATION

At the international level, protocols have been agreed upon that implicitly promote the harmonization of regulatory systems. While the Codex *Principles for the risk analysis of foods derived from modern biotechnology* are available to guide the safety assessment of GM food, they have no binding effect on national legislation, but do form the basis for harmonization under the SPS Agreement. On the other hand, the CPB has established legally binding rules for environmental risk assessments.

In addition, OECD has experience in promoting international harmonization in the regulation of biotechnology by ensuring efficiency in the evaluation of environmental and human-health safety, through its working group for harmonization in biotechnology and its task force for the safety of novel foods and feeds. Developing countries therefore have sets of agreed principles for guidance, and the advantage of learning from the experiences of their forerunners by investigating best practices and adapting them to suit their individual situations. Although agreement has been reached on the scientific principles of food-safety assessment, consensus has not been achieved on the extent of data required to comply with these principles or on the role of the data in decision-making.

Harmonization of components of the scientific review process has a potential benefit where a lack of resources threatens effectiveness, and the affected countries in the region have determined and agreed on the regulatory objectives.

The advantages of regional/subregional cooperation are to facilitate regulation, promote the sharing of resources, synchronize the assessment of foods derived from modern biotechnology, and expedite information exchange. The Nuffield Council of Bioethics recommends the implementation of international standards and the sharing of risk-assessment methodologies and results, particularly between developing countries with similar ecological environments.

Moreover, integrating some activities could reduce the overall requirement for new financial resources. Harmonization can be achieved at several levels, *i.e.* some elements of the framework can be implemented at the regional level. The countries of the Association of South-East Asian Nations have come together to cooperate on various levels, including:

- Harmonization of legislation for products derived from modern biotechnology and intellectual property rights;
- RandD in biotechnology; and
- Environmental protection.

ASEAN is also looking at a regional approach to biosafety, although it is not clear what is intended, *i.e.* whether regional assessment and national decision-making would be considered. Those countries in the region that have made some progress have gone as far as developing labelling regulations, although they acknowledge that implementation may not be possible in the near future due to a lack of human resources.

After the 2002 humanitarian crisis in southern Africa, where a number of countries experiencing severe drought and food shortages questioned the use and safety of GM food aid, a Council of Ministers of the Southern African Development Community established an Advisory Committee on Biosafety and Biotechnology to develop a common position on biotechnology and harmonize biosafety legislation in the region.

The objective is to facilitate the movement of food products that may contain GM material across the region in future. Although harmonization may absorb some of the costs that could be incurred in establishing regulatory frameworks, the flexibility allowed by international agreements creates room for divergence from the basic principles. Also, none of the regimes give guidance on regulations. Therefore, achieving harmonization in this context may be debatable as countries grapple with criteria for the precautionary approach and socioeconomics. Nevertheless, particular attention should be paid to supporting and creating new strategic partnerships. Countries need to find effective ways of working together, and analyse the benefits and costs of harmonization.

AGRICULTURAL BIOTECHNOLOGY AND FOOD SAFETY

When a consumer buys food s/he expects it to be safe and have the same quality characteristics that s/he is accustomed to, and familiar with. Consumers have the right to know that this is the case and if there is anything different about that product.

The Food and Agriculture Organization and the World Health Organization of the United Nations advocate the concept of 'substantial equivalence' as the most practical approach to address the safety evaluation of foods or food components derived through modern biotechnology.

This approach states that 'if a new food or food component is found to be substantially equivalent to an existing food or food component, it can be treated in the same manner with respect to safety.' So what kind of tests are used to determine if the food is as safe or safer than existing foodstuffs? Researchers must prepare comprehensive data to support the safety and wholesomeness of new crop varieties developed through biotechnology.

This process requires years of labouratory and field testing before a product can be brought to the market. To provide assurance that foods derived through biotechnology are as safe as those produced by traditional breeding programmes, the safety assessment strategies involve several key steps. These steps include molecular characterisation of the genetic modification, agronomic characterisation, nutritional assessment, toxicological assessment and safety assessment.

For example, typical questions that must be addressed are:

- Does the genetically modified food have a traditional counterpart that has a history of safe use?
- Has the concentration of any naturally occurring toxins or allergens in the food changed?
- Have the levels of key nutrients changed?
- Do new substances in the genetically modified food have a history of safe use?
- Has the food's digestibility been affected?
- Has the food been produced using accepted, established procedures?

Even after these and other questions about the biotechnology derived foods are answered, there are still more steps in the approval process before the crop can be commercialized. In fact, genetically modified foods are the most studied food products ever produced.

- *Safety assessment*: Molecular characterisation – for new plant varieties produced through modern biotechnology, the source of the gene introduced into the plant is first identified. The transformation system used to insert the gene is defined, as well as the number of copies of inserted genes and the integrity and stability of the genetic insert.
- *Agronomic traits*: Usually the starting points for evaluating substantial equivalence. For example, in the case of potatoes, the traits commonly examined are yield, tuber size and distribution, dry matter content and disease resistance.
- *Nutritional assessment*: Involves key nutrients including fats, proteins, carbohydrates and essential vitamins and minerals.
- *Toxicology assessment*: Toxicants and anti-nutrients are compounds known to be naturally present in some crops that could have an impact on health if their levels increased. For example, solanine glycoalkaloids in potatoes or trypsin inhibitors in soybeans). The levels of anti-nutrients in crops produced through biotechnology are compared to conventionally produced varieties grown under comparable environmental and agronomic conditions.
- *Safety assessment*: When a crop produced through biotechnology is shown to be substantially equivalent to a conventional crop, the safety assessment focuses on the introduced trait and the expressed protein product.

The overall goal of these tests is to determine whether the plant is substantially equivalent to food derived from a conventional source that has a history of safe use. A substantial equivalence evaluation focuses on the product rather than the process used to develop the product. If the new product is substantially equivalent to the conventional food or feed, then the product derived through biotechnology is considered to be as safe as the conventional counterpart. If the food produced using biotechnology contains a new trait, which changes the levels of nutrients or antinutrients, such as a higher level of a vitamin or a lower level of an allergen, the assessment focuses on demonstrating the safety of the new trait.

ALLERGENS

One of the public's biggest concerns related to genetically modified foods is that an allergen could be accidentally introduced into a food product. The DNA inserted into a plant is safe and no case of allergy to DNA is known, the concern is whether the protein produced by the plant as a result of the inserted DNA could cause an unexpected allergic reaction. Scientists know a lot about which foods trigger allergic reactions in adults and children. Ninety per cent of all food allergies are associated with only eight foods or food groups – shellfish, eggs, fish, milk, peanuts, soybeans, tree nuts, and wheat. These, and many other food allergens, are well characterized. To date, genes found to express any known allergenic protein have been specifically excluded from the research and development process. Scientists have agreed that following these simple steps provides compelling assurance that no allergic effects will result from foods derived from biotechnology:

- Avoid transferring genes from foods known to be allergenic.
- Check the structure of any new proteins produced in foods derived from biotechnology against the structures of known allergens to assure that no allergenic structures exist in the new protein.
- Measure the stability of the new protein in stomach and intestinal fluids. Most allergens are stable to these conditions. Proteins which are unstable to these conditions are not likely to be allergens.
- Determine how much of the new protein will be present in the food consumed by humans. Most allergens are present in large amounts.

Allergenicity screening continues to be a very important part of safety testing before a crop can enter into the food market.

ANTIBIOTIC RESISTANCE

Some biotechnology crops contain genes for a trait called antibiotic resistance. Scientists use this trait as a marker to identify cells into which the desired gene has been successfully introduced.

Concerns have been raised that these marker genes could move from biotechnology crops to microorganisms that normally reside in a person's gut and lead to an increase in antibiotic resistance. There have been numerous scientific reviews and experimental studies of this issue and they have come to the following conclusions:

- The likelihood of antibiotic resistance genes moving from biotechnology crops to other organisms is extremely remote.
- In the unlikely event that an antibiotic resistance gene is transferred to another organism, the impact of this transfer would be negligible, as the markers used in genetically modified crops have limited or no clinical or veterinary use.

However, in response to public concerns, scientists have been advised to avoid using antibiotic resistance genes in biotechnology modified plants. Alternative marker strategies are being developed.

During these first decades, the biophysical changes as such will be less pronounced but climate change will affect those particularly adversely that are still more dependent on agriculture and have lower overall incomes to cope with the impacts of climate change. By contrast, the second half of the century is expected to bring more severe biophysical impacts but also a greater ability to cope with them. The underlying assumption is that the general transition in the income formation away from agriculture towards non-agriculture will be successful.

How strong the impacts of climate change will be felt over all decades will crucially depend on the future policy environment for the poor. Freer trade can help to improve access to international supplies; investments in transportation and communication infrastructure will help provide secure and timely local deliveries; irrigation, a promotion of sustainable agricultural practices, and continued technological progress can play a crucial role in providing steady local and international supplies under climate change.

Bibliography

A S Karwa; M K Rai and H B Singh: *Handbook of Techniques in Microbiology : A Laboratory Guide to Microbes*, Scientific Publication, Jaipur, 2008.

A. Poshadri and Aparna Kuna : *A Handbook of Food Techno's*, New India Publishing Agency, Delhi, 2013.

A.K. Shrivastava : *Agriculture and Food*, APH Publication, Delhi, 2004.

Anubhav Dwivedi : *A Fragrance from Food Production World*, Axis Publication, Delhi, 2010.

Archana Ruhela : *Agriculture and Food Security*, Oxford Book Company, Delhi, 2008.

Arnold C. Long : *Advances in Seafood Biochemistry: Composition and Quality*, Cyber Tech Publication, Delhi, 2009.

B. K. Mishra: *Dairy and Food Processing Industry : Recent Trends (2 Vols-Set)*, Biotech Books, New Delhi, 2014.

B.D. Sharma.: *Agriculture Development and Food Security*, Ancient Publishing House, Delhi, 2012.

Boria Majumdar : *Cooking on the Run : An Average Indian Man's Encounters with Food*, Harper Collins, Delhi, 2012.

Bruno Dorin and Frederic Landy : *Agriculture and Food in India: A Half-Century Review*, Manohar Publication, Delhi, 2009.

C S Jain: *A Complete Book on Health and Nutrition*, Cyber Tech Publications, Delhi, 2009.

Dilip Shah: *Co-Operativization Liberalization and Dairy Industry in India*, ABD Publishers, Delhi, 2000.

Doreen Virtue: *Constant Craving*: *What Your Food Cravings Mean and How To Overcome Them*, Hay House India, 2011.

Ernest R. Vieira.: *Elementary Food Science*, Chapman and Hall, 2010.

Eugene Lyman Fisk, Adelle Davis, Florence Daniel, Ruth A. Wardall and Harry Snyder.: *Food Science and Nutrition* (*Vols* 1 *to* 2 *Set*), Shree Publications, Delhi, 2009.

Harmeet Singh: *Dairy Farming*, APH Publication, Delhi, 2011.

Hema Thapar: *Food Science and Health*, Pacific Publications, Delhi, 2011.

Jerry D Souza and Jatin Pradhan: *Handbook of Food Science Catering Technology and Kitchen Management*, SBS Publications, Delhi, 2010.

M Lakshmi Narasaiah: *Economic Growth and Food Security*, Discovery Publishing House, Delhi, 2008.

Manoj Negi: *Career in Dairy Farming*, Abhishek Publication, Delhi, 2010.

Mohammad Raziuddin and Ashok Hembade: *Milk and Milk Products Technology*, Jaya Publishing House, Delhi, 2013.

P.C. Trivedi: *Microbes : Applications and Effects*, Aavishkar Publication, Delhi, 2009.

Ravee Chauhan: *Advanced Hotel Industry and Tourism*, Vista International Publishing House, 2011.

Renu Singh: *Food and Beverage Management*, Delhi, Rajat Publications, 2012.

S. Sundara Rajan: *Genetics and Physiology of Microbes*, Anmol Publication, 2003.

S.K. Soni: *Microbes : A Source of Energy for 21 Century*, New India Publication, Delhi, 2007.

S.K. Tomar, R. Singh, A.K. Singh, Sumit Arora and R.R.B Singh: *Functional Dairy Foods Concepts and Applications*, Satish Serial Publication, Delhi, 2011.

Udai Veer: *Elements of Food Science*, Anmol Publications, Delhi, 2007.

W. Merback and A. Veha : *Crop Science and Technology for Food Security Bioenergy and Sustainability*, L. Bona, J. Pauk, Agrobios Publication, Delhi, 2012.

W.L. Slatter, T. Krishtofferson and D.A. Seiberling: *Manual of Dairy Industry : Training and Development of Personnel for the Dairy Industry*, Asiatic Publication, Delhi, 2012.

Index